MAP AND TOP

Books are to be returned on or before
the last date below.

MAP and TOP
ADVANCED MANUFACTURING COMMUNICATIONS

John Dwyer and Adrian Ioannou

Foreword by W. John Eichler,
General Motors Corporation

KOGAN PAGE

Dedication

To the poor bloody infantry of manufacturing industry, high and low, on whom the rest of the ingrate world relies, this book is dedicated.

First published in Great Britain in 1987
by Kogan Page Ltd, 120 Pentonville Road, London N1 9JN

Copyright © John Dwyer and Adrian Ioannou 1987
All rights reserved

British Library Cataloguing in Publication Data

Dwyer, John
  MAP and TOP : advanced manufacturing communications.
  1. Control theory
  I. Title   II. Ioannou, Adrian
  629.8   Q4402.3

ISBN 1 85091 355 2

Photoset in North Wales by
Derek Doyle & Associates, Mold, Clwyd
Printed and bound in Great Britain by
Richard Clay, The Chaucer Press, Bungay

CONTENTS

FOREWORD by W. John Eichler, General Motors Corporation   ix
PREFACE by J.A. Collins, OBE   xi
ACKNOWLEDGEMENTS   xiii

1   INTRODUCTION   1

2   CURRENT MANUFACTURING TECHNIQUES   13

3   THE NETWORK – THE PHYSICAL TRANSMISSION MEDIUM   23

    3.1   Basic communications 23

    3.2   Communications options 25
        3.2.1   *Twisted pair* 26
        3.2.2   *Common connecting methods* 27
        3.2.3   *Other twisted-pair systems* 29

    3.3   Distributed LANs 30
        3.3.1   *Fibre optics* 33
        3.3.2   *Ethernet* 35
        3.3.3   *Token passing* 38
        3.3.4   *Broadband and baseband* 41

    3.4   Head ends 44

4   THE SEVEN LAYER MODEL   47

    4.1   The application itself 53

    4.2   Applications layer (layer 7) 53
        4.2.1   *ACSE, formerly called CASE* 54
        4.2.2   *FTAM* 55
        4.2.3   *MMFS and MMS* 56
        4.2.4   *Other SASE elements* 59

    4.3   Presentation layer (layer 6) 62

    4.4   Session layer (layer 5) 62

4.5 Transport layer (layer 4) 63
    4.5.1 *Connection-oriented and connectionless communications* 63
    4.5.2 *Error correction* 66

4.6 Network layer (layer 3) 69

4.7 Datalink layer (layer 2) 70

4.8 Physical layer (layer 1) 71
    4.8.1 *Frame lengths* 72
    4.8.2 *Carrier band* 73
    4.8.3 *Enhanced Performance Architecture (EPA)* 76
    4.8.4 *Why bother with broadband?* 78
    4.8.5 *Field bus* 79

4.9 Interconnecting MAP and TOP with each other and with other systems 81

4.10 Technical and Office Protocols (TOP) 84

5 THE STANDARDS SCENE   91

5.1 The origins of OSI 91

5.2 The International Standards Organization (ISO) 92

5.3 ISO stages of development of standards 95

5.4 National activity in the UK 97

5.5 Who's who in standards 98

6 KEY ISSUES   101

6.1 The functionality of MMS 101
    6.1.1 *CNMA's concerns* 105
    6.1.2 *Migration* 106
    6.1.3 *Companion standards* 107
    6.1.4 *Ambiguities* 107

6.2 Interoperability 109

6.3 Testing 109

7 MAP AND TOP APPLICATION CASE STUDIES   115

7.1 The Towers of Hanoi and beyond 115
    7.1.1 *Autofact 1985* 115
    7.1.2 *Lessons learnt* 118
    7.1.3 *The way ahead* 119

7.2 Cell controller applications 119
    7.2.1 *Application 1* 121

    7.2.2 *Application 2* 122

  7.3 A link from design to assembly and inspection 124

  7.4 MAP in the electronics test environment 127

  7.5 AIMS – an Assembly Information Management System applied to engine assembly 133
    7.5.1 *The solution* 136
    7.5.2 *Autonomous operation* 138
    7.5.3 *Real-time response* 138
    7.5.4 *Standardization* 139
    7.5.5 *The future* 139

  7.6 Using MAP in the factory 139
    7.6.1 *MAP implementation* 139
    7.6.2 *MAP network access methods* 142
    7.6.3 *Data flow and message traffic* 144
    7.6.4 *MMFS Robot Class 2 – typical messages* 146

  7.7 MAP in printed circuit board assembly 148

  7.8 Communications in the aerospace industry 151

  7.9 MAP in General Motors 163
    7.9.1 *Saginaw* 163
    7.9.2 *The GMT400 Truck and Bus Project* 167

8 THE WAY AHEAD   171

  8.1 The future 171
  8.2 MAP and TOP products 172

USEFUL ADDRESSES   175

ABBREVIATIONS   179

SOURCES AND RECOMMENDATIONS FOR FURTHER READING   187

INDEX   191

WHY NETWORKING STANDARDS?

Half the cost of automating factories is in supporting multiple proprietory networks and redundant plant wiring.

GM MAP Task Force, 1981

Foreword

Advances in technology are making the business and manufacturing environment increasingly complex. Standards can help us cope with this complexity. Given the strategic importance of computers in the economies of the industrial world, it is fitting that one of the most significant commercial stories of our time is the standardization of computer communications. Quite frankly, when we joined with other computer users to launch this effort we didn't predict its scope and public visibility. In restrospect, I guess we should have done.

The computer assisted technologies looming on the horizon offer some of the greatest functional and productivity tools available to improve business operations. However, the absence of a standardized electronic link permeating most business organizations poses a severe impediment to the efficient deployment of this technology.

The feasibility of using computer controlled devices to design, test, and manufacture products – as part of a massive network – is well within our technological grasp. However, unless the world agrees upon a global set of standards that will make multi-vendor computer systems interoperable, successful implementation of these technologies becomes less and less attractive.

Because this goes to the heart of our business and the long-term strategic thrust of General Motors, we vigorously endorse the international call for recognition of the Manufacturing Automation Protocol and Technical and Office Protocols for, respectively, the plant floor and office. Obviously, we are not alone. To date, more than 1,400 companies have joined the international federation of MAP/TOP Users Groups in Australia, Europe, Japan and North America.

There is no question that the worldwide MAP/TOP community is pleased with the progress to date. However, they are impatient to complete the task; as the authors point out, many things remain to be accomplished. Fortunately, the growing support and the team work being demonstrated between users, suppliers, standards groups, and other organizations assures success. Soon users all over the world will be benefiting from the successful implementation of these standards.

When that does happen it will mean that standardization will be doing for the computer industry what it has done for my industry – autos. Standardization of parts and components permitted the automobile to be the backbone of the industrial world and standardized locations for brake, clutch, and accelerator pedals have allowed any driver to operate any vehicle without major retraining. As with autos, standardization has the potential of placing powerful computer tools in the hands of a larger number of people – not just a privileged few.

In the end we all benefit. Computer suppliers broaden their markets and users become more productive.

When MAP/TOP does become universally accepted it will be due in no small part to the exchange of information among people everywhere trying to come to grips with the same problems. Because of this, the MAP/TOP community is indebted to the authors, John Dwyer and Adrian Ioannou. Their book will help lend credence to the statement that an informed user is always the wisest consumer.

W. John Eichler
Director
Advanced Engineering Staff
General Motors Corporation

PREFACE

The last two decades have witnessed fundamental changes in the application of automation to the manufacturing industry. Much of this change can be attributed to the improvement in the mechanisms of manufacturing processes, but by far the greatest influence has been from the application of the computer to the control of the equipment and processes involved.

In its simplest form this has converted what was essentially a direct man/machine relationship to a 'hands-off' situation, resulting in the ability to program and control complex processes either in real time or off-line. From a start point of a single machine the same degree of control this computing application gives has been extended to the immediate equipment linking similar machines and in so doing has created an island of automation.

In parallel with these advances, the application of the computer has substantially altered the control procedures used to program a whole manufacturing process and linked them to the commercial functions from marketing, through design, to sales and distribution. Thus we have arrived at a position where the possibilities and advantages of computer integrated manufacture (CIM) can clearly be seen to be both real and viable. This vision has greatly stimulated both engineers and administrators to tackle the most formidable remaining obstacle to the successful introduction of CIM, namely effective communication.

As with all processes of change which are achieved by degree, the advances have been made by separate technological steps from companies, or even individuals, engaged in similar, though not identical, operational methods. One of the main considerations of such work is the possibility presented of the commercial advantage of creating a universal system which ensures its widest adoption – a consideration which is rarely, if ever, achieved.

The consequent lack of standards has led to the present situation in which it is difficult and expensive to link similar equipment and systems which has, in turn, led to the frustration of those wishing to benefit from CIM.

The manufacturing automation protocol (MAP) initiative of General Motors Corporation of the USA is the most comprehensive proposal to resolve these issues on the shopfloor and, combined with technical and office protocol (TOP), covering the associated office operations, offers a complete approach to the problems throughout the manufacturing environment.

This volume is dedicated to a complete analysis, at all levels, of the nature of MAP and TOP, their application, the techniques used, the problems involved and their solution. It is to be recommended to all who are interested in the subject of CIM as a guide and reference work to one of the most stimulating areas of new thinking in the field of manufacturing engineering.

For those who believe that manufacturing has a future and that it lies in the successful harnessing of the power of the computer, MAP and TOP open the way for this to be achieved and the authors are to be commended for their timely and erudite work in assisting this vital step.

J A Collins, OBE
Group Director,
Manufacturing Engineering,
TI Group PLC

Acknowledgements

This book could not have been written without the help of a large number of people, many of them outstandingly generous with their time and patience in filling in yawning gaps in the authors' knowledge. In particular the authors would like to thank:

Jack Eichler, Mike Kaminsky, Ron Floyd, Gary Workman, John Tomlinson, Mark Adler and Mark Cocroft of General Motors
Jim Doar of Boeing
Brian Phillipson and Malcolm Golding of British Aerospace
Peter Cornwell of Renishaw Controls
James deRaeve, Tony Rixon and Ken Beese of The Networking Centre
Ken Olsen, Paul Evans, Bob Grindley, Dennis LaBarge and Wayne Adams of Digital Equipment Corporation
Tony Domenico, John Lord and others of Electronic Data Systems
Len Magnuson and Vijay Kochar of Intel
Peter Crooks, James Langdal, Dedy Saban and Rhonda Dirvin of Motorola and Clive Gay of Motorola (UK) Ltd
Jim Kenny, Dave Thompson and Roy Cadwallader of ICL
Ken Spenser, Dan Moon and Alleyne Leach of Texas Instruments, UK
Colin Hoptroff of Jaguar Cars and the European MAP Users Group
Carl Clarke of Leeds University
Graeme Wood of Foxboro
Colin Pye of the National Computing Centre of the UK
Tony Helies and Suzanne Barclay of Concord Communications
Joe Schoendorf of Industrial Networking Inc
Chuck Gardner of Eastman Kodak
Vick Gregory and Gerry Brocklebank of Unilever
Richard Bleasdale, Graham Symcox and Andrew Gray of Coopers and Lybrand
Bob Crowder of Ship Star Associates
Nick Beale
Dick Lefkon of Citicorp
Andy McMillan and Bob Digiovani of the Industrial Technology Institute
Paul Cheshire of CAP
Nigel Kingsley of Mars Electronics

Dennis Mottram, David Beaumont and Dr John Beeston of GEC of the UK
Frank McKenna of British Petroleum
Jack Klein, Mike Sutton, Phil Evins and Don Ralston of IBM
Peter Baker and Peter Burton of ERA Technology
Allan Eaton, Gerry Hibbs, Don Wilkin and John Crane of Allen-Bradley
Steve Yeardley and Martyn R. Jones of Gould Electronics
Dr Peter Summerfield and Ivan Wheatley of Rover Group
Mike Grant, Peter Marchant and Ron Yeomans of Istel Automation
Mike Nichol and Alan Fitzgerald of Scicon
Dave Blanchard of Hewlett-Packard (USA)
David Pearse and David Burnage of Reflex Automated Systems & Controls Ltd
John A. Collins, OBE of Tube Investments Group Plc
Paul Drummond and Robert B. Michaelson of P-E Consulting Services Ltd
Roger J. Gibbons of the British Standards Institution
Dave Lyon of Hewlett-Packard (UK)
Paul MacAree of Tandem Computers (UK) Ltd
Andy Poupart of Tandem Computers Inc
Archie Scott and Anthony Wood-Smith of Honeywell (UK) Ltd

And a special thanks to the heroes and heroines at the Department of Trade and Industry who dreamed up and organized CIMAP: Peter Adkin, Dr David Miles, Dr Pauline Curtis and Ron Downing and their collaborators Richard Knight of Findlay Publications and Ron Armstrong and his colleagues at PERA.

The insights in this book are attributable to all these people; any mistakes are ours alone.

1
INTRODUCTION

There are many reasons why the Manufacturing Automation Protocol (MAP) first drawn up by General Motors and the Technical and Office Protocols (TOP) promoted by Boeing are important. One is that, in the short but turbulent history of the computer industry, these two initiatives mark the first occasion on which users dictated terms to even the mightiest of their suppliers. Another is this: MAP and TOP specifications were drawn up for use by those in manufacturing industry, and this book is intended mainly for these users. But MAP and, especially, TOP may have much wider applications than in manufacturing. The authors hope readers in other activities than manufacturing will also find the book useful and relevant, despite the inevitable preponderance of references to 'manufacturing' in the text.

Nevertheless a main aim of this book is to help every manufacturing manager undertake one of the most urgent of the many tasks he or she is faced with today. That task is to become familiar with at least some of the theory and practice, first, of modern communications technology in general and, second, of MAP and TOP in particular.

MAP is not the only factory-communications system that will be available and many other office networks besides TOP are already around. Some, like RS232, Ethernet and the like, are already familiar in both factory and office. There is talk of others, known in the standards community as 'functional standards', which could become rivals to MAP and TOP. But MAP, in particular, already has so much clout behind it that the other functional standards are unlikely to succeed in doing much more than delay – perhaps damagingly – the inevitable universal acceptance of MAP, perhaps in a more highly developed form, as the factory communications system for the year 2000.

True, enormous problems remain before MAP or TOP can be taken for granted. Some of these, such as conformance testing, are explained in the book. Potential users have to take this into account. But MAP and TOP, regardless of their current limitations, seem to offer one vital component in manufacturing's strategy for survival. Survival, indeed, is what manufacturing is about today. Survival means aiming for global markets or, in the jargon, becoming world-class suppliers. It means the lowest cost production consistent, always, with the highest quality.

MAP and TOP will not, by themselves, enable users to achieve these things. But MAP and TOP are technologies which a company's competitors can use to help those competitors achieve *their* goals. This means that any company – whatever its size – that is unacquainted with the power of effective inter-company and intra-company communications will be at a disadvantage. The authors of this book are independent and have no stake in the success or failure of MAP other than as beneficiaries or victims of its effects on manufacturing industry. But it seems clear to us that the companies that survive, and thrive, in the next century will be those which have taken on board, and thoroughly understood, the implications of modern communications technology. It is sad but inevitable that the rest will not be around to contest the argument.

Moreover the most successful users of the technology will be those who acquire the communications skills themselves, rather than relying on vendors to supply the skills for them. In the future every well-run factory will have in-house communications expertise on hand to function much as today's electricians do. There is no need to become a communications expert to acquire enough knowledge of the subject to allow informed choices among the equipment and systems which rival vendors will offer. This book has been written to fill the immediate need for this basic knowledge.

So what are MAP and TOP? Both were born of irritation with the need to provide a separate communications network from every vendor who sold a manufacturing system. This meant new, and expensive, cabling every time a CADCAM system, robot cell or data collection system was brought into the factory. What is more, none of these separate networks could be interlinked with any of the others.

In the case of MAP, General Motors had ambitious plans to automate its factories. In 1985 it spent $9 billion on new plant and planned to spend $35 billion in the five years to 1990, when it would have had 200,000 programmable devices in its various plants. This is five times the number at the end of 1984. But, by 1980, GM had already grown tired of robot, conveyor and machine-tool controllers and shop-floor terminals that didn't talk to each other. Parts of production lines linked back to controlling screens and keyboards in the control room in such a way that supervisors were faced with up to five sets of screens and keyboards from different suppliers to oversee the operation of a single part of the production process. Information which came up on one screen would have to be keyed into another. For example, GM uses IBM machines to keep a record of who is at work on any particular day but it uses Digital Equipment (DEC) computers to schedule the work to be done on a particular day. The two sets of information have to be combined to produce a work schedule for the day and the only way to do it is to read and re-keyboard. Again, a Hewlett-Packard machine may be acting as a cell controller. If it reports that there is a malfunction because of a worn part, the supervisor has to refer to an IBM machine to find out if there is a replacement part and a third machine to find out who can do the repair. Only 15 per cent of the 40,000-plus devices on GM's shop floor in 1984 communicated beyond their own processes, and 40 per cent of GM's investment went into communications, much of it wasted in supplying interfaces to sort out this computerized Babel.

In any one factory automation project, somewhere between a third and a half of the total cost may be incurred in the wiring and software needed to get the various parts, whether robots, programmable controllers or computer numerical control (CNC) machines, to talk to one another. According to GM's speakers at various MAP conferences, if the various communications systems conformed to some sort of standard the saving in a project worth $20 million would be $2 million to $3 million. Early in 1986 GM said it expected to install 155,000 computer-controlled machines by 1990. Using conventional methods the cabling costs and special software would run to $1,600 for each of these machines. But GM has calculated that, using standard interfaces, this could go down to $300 each, saving $201.5 million. This saving is achieved through not having to write new communications software to link part A, whose communications system speaks one language, with part B, whose communications system speaks another. This is independent of the need to write interfaces for the applications software running between those devices.

The factory automation industry had no standards so, in 1980, GM set up a task force to see if standardization was possible. MAP, which first emerged in 1982 and was adopted by GM in October of that year, is the result.

GM's idea was to use a single MAP broadband cable for all a factory's communications traffic. Everything – every programmable controller, robot controller, machine tool controller and factory floor terminal – would plug into it and communicate, on demand, with everything else on the cable. The cable would also connect any of these systems, if needed, to factory-wide mainframe manufacturing control systems. This would make shop-floor supervision and factory control much easier. Order entry systems could link directly into inventory control and scheduling systems. Schedules could be downloaded straight into shop-floor terminals and could trigger the changing of control programs on individual machines, if necessary. There was no limit to what could be achieved, given enough imagination by management. And, as explained more fully in the next chapter, the indirect benefits would probably outweigh those that were immediately apparent.

Boeing's problems were similar to those of GM. The average modern jetliner is made up of 3.5 million parts, each of which needs a drawing and the associated process data which explains how to make the part. Boeing wanted a means of exchanging this data within Boeing's factories, within its offices and between its offices and factories. It seized immediately on GM's MAP initiative and adapted MAP to its own needs. The Technical and Office Protocols (TOP) is the result.

GM told its suppliers they would get no more sales unless they conformed to MAP and Boeing issued a similar ultimatum to its suppliers. GM's suppliers were particularly anxious to comply. GM is the world's largest company. Its sales in 1986 were about $104 billion, and it has a corresponding amount of buying power. Boeing is one-tenth the size of GM but 1985 sales of $10 billion were still persuasive. The two together looked unstoppable.

All the same GM, as the instigator of the MAP programme, knew it had to offer its suppliers something in return for their co-operation in developing as-yet non-existent communications systems. It turned MAP into a worldwide crusade. GM insisted that MAP conform to internationally agreed standards. This meant using the International Standards Organization's (ISO's) Open Systems Interconnect (OSI) communications model which itself became an international standard in the late seventies. This meant the vendors of MAP (and, by extension, its derivative, TOP) would have potential sales in factories and offices all over the world.

On 24 April 1984, GM, Boeing, the late US Commerce Secretary Malcolm Baldrige and the director of the National Bureau of Standards in the USA, Ernest Ambler, signed an agreement committing the four parties to a programme of tests and demonstrations to implement the selected parts of the OSI of which MAP and TOP would consist. The first such demonstration was to be held at the National Computer Conference in Las Vegas, Nevada, in July that year. In November 1985, 20 vendors collaborated to demonstrate a MAP network making plastic models at the Autofact exhibition in Detroit. The bandwagon had begun to roll. The next big demonstration, CIMAP (December 1986), was organized without much help from either GM or Boeing. At CIMAP over 60 companies came together in Birmingham, England, in

13 demonstrations on MAP and TOP networks. The demonstrations were arranged by the UK's Department of Trade and Industry and were the most ambitious to date, mainly because many of them used actual parts and sub-assemblies used in the car and other industries instead of plastic toys. Figure 1.1 shows CIMAP's event network layout.

The final demonstration in the series, called Enterprise Network Event, is set for Baltimore in June, 1988. One of the conditions the US MAP/TOP Users' Group has laid down for participation in this event is that all the communications systems and interfaces demonstrated must be of commercially available products. If the conformance test timetable is kept, which is problematic, these products will conform to MAP and TOP version three, and not, at least in MAP's case, the 2.X development versions that, so far, have been all that is available. This is an important milestone for users and suppliers. It will mark MAP's and TOP's transition from idea to reality. If Enterprise Network succeeds, this does not mean there is no more work to do, but it means MAP and TOP have reached some kind of near-maturity.

The result of all this activity is this. A few years ago hardware and software suppliers could not have cared less about the inability of their equipment to talk to anything else, on or off the shop floor. Now they are falling over themselves to emphasize how compatible their systems are with everyone else's. It is probably the most important change – and the biggest challenge – since the arrival of the data-processing department.

Demonstrations aside, how far has MAP got? No-one yet knows how well a full up-and-running MAP system is going to perform, though we are likely to have a much better idea once the several MAP pilots now being installed are operational. One such pilot is GM's Steering Gear Division plant at Saginaw, Michigan (Figure 1.2). At Saginaw the plan is to link up about 40 manufacturing cells containing some 60 robots in a 70,000 square foot plant, one of five at the site. Saginaw is a laboratory for MAP-connected systems for the rest of GM. GM's plans have been revised somewhat but GM's intention remains to spend $52 million to set up a completely integrated, paperless factory at Saginaw. The plant will make three components for a front-wheel-drive axle and assemble them with about 20 other components into a finished axle. GM has other plants making these axles – Saginaw will make only 10 per cent of GM's needs – so GM will, when the factory is complete, and integrated with its CADCAM systems some time in 1988, provide direct comparisons of the cost, quality, delivery and reliability between products made conventionally and products made by automated manufacturing.

6 MAP and TOP: Advanced Manufacturing Communications

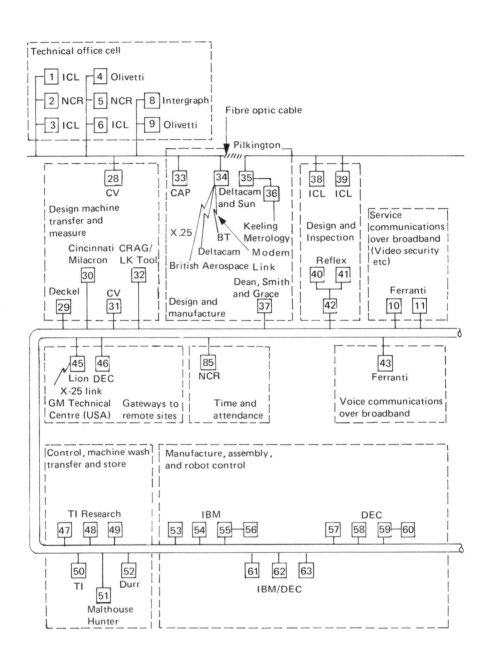

Figure 1.1 CIMAP event network layout

Introduction 7

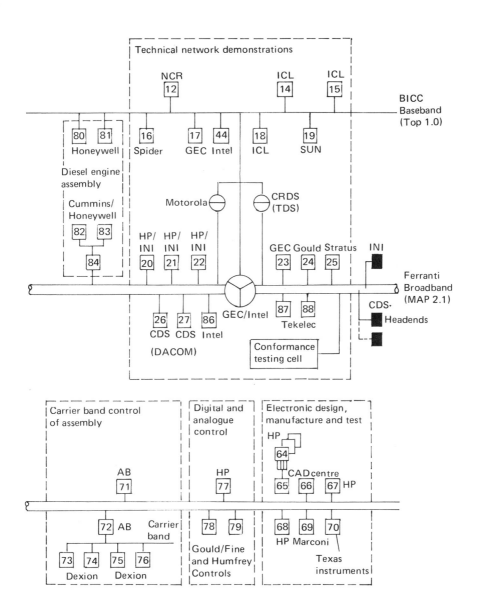

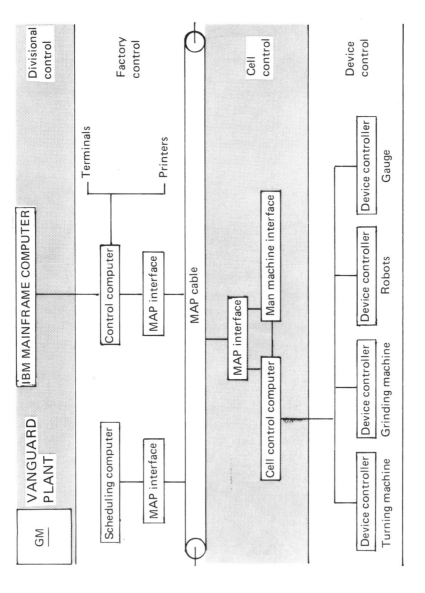

Figure 1.2 General Motor's Vanguard plant at Saginaw, Michigan

The other main focus of interest is GM's truck and bus group, five of whose factories were running MAP networks by the end of 1986. Some $25 million of the cost of this $350 million 'GMT400' project is MAP-based communications. Another project, GM10, involves MAP-connecting two or three car sites. Altogether, about 30 GM plants in North America will be using some MAP communications network by mid-1988.

GM admits to having had technological problems with some of these plants and, though some of the problems appear to relate more to lack of training than to MAP, some of the more interesting implementations may turn out to be those outside GM.

In the USA MAP's users cut a swathe right across American industry. John Deere of the USA is putting in systems. IBM, an early adviser to GM on MAP, has installed a pilot MAP system at its Endicott, New York factory, and the Kaiser Aluminium & Chemical Corporation aims to be the first heavy process company to put in a system, to produce forgings. Du Pont is putting in a 2,000 ft MAP cable to make breech parts for sporting rifles. Eastman Kodak is also installing a system. Even companies once sceptical of the concept appear to be falling into line. Du Pont, like GM itself, now adopts the attitude that anyone who wants to sell them equipment can either ignore these standards-based specifications or supply Du Pont – they can't do both. And note that Du Pont is a process company, not a piece part manufacturer. Proway, a process-control communications standard (see Section 4.8.2), has now been written into MAP, in version 2.2. In Europe the process control companies tend to be more active participants in MAP activity than their counterparts in the USA. The first chairman of the European MAP Users Group (EMUG) was Vic Gregory of Unilever.

EMUG now has 200 members. European companies supporting MAP now include Siemens and ASEA, to name but two, and the present chairman of the European MAP Users Group is Colin Hoptroff of Jaguar Cars of the UK. Jaguar is installing a MAP pilot, as is the Rover Group. Also in the UK, computer company ICL was involved in the Las Vegas demonstration and is in the early stages of putting in a four-level MAP-based communications system. After an initial period in which the European MAP and TOP users groups were combined, a separate TOP users' group was formed in Europe to make sure the interests of the financial community were not overridden by manufacturing concerns. And the European Commission's ESPRIT programme has funded two communications projects, 955 and 688, which use and complement MAP, though these were put at risk early in

1987 by the refusal of the UK to approve the framework budget which funds ESPRIT and other EEC sponsored research.

Neither MAP nor TOP is a cure-all. The development of skills in communications technology is necessary, but it is not the sole condition for manufacturing success. And MAP in its present state could be considered as far from satisfactory. It is doubtful, for example, whether any of GM's implementations listed above could talk to each other, however well they work on their own. The reason is that the standards on which MAP is based were largely unwritten four years ago but, in the meantime, General Motors and its collaborators had to get to work. They filled the gaps in the existing standards wherever they could. An example of this is a protocol called the Manufacturing Message Format Standard (MMFS, pronounced 'Memphis').

For MAP's first demonstration at the 1984 National Computer Conference in Las Vegas the seven participants were linked over the first version of MAP, MAP 1.0. The Saginaw Steering system uses a variant of MAP 2.0 which is probably unlike any other MAP version because Saginaw is continually migrating towards the latest MAP specifications. And in theory both the Autofact show in November and the CIMAP show in England used MAP 2.1, though the MAP version used for Autofact 85 varied slightly from strict MAP 2.1. The version used at CIMAP was in some respects the same as the Autofact 85 version. Some of these MAP variants differed in such things as message format or running speed – 5 or 10 megabits a second – though all new MAP systems are now standardized on 10 megabits a second.

Another source of incompatibility was that MAP interfaces from the same supplier worked well with each other but not with the interfaces from a different supplier. This embarrassing failing – MAP after all is meant to allow combinations of different vendors' products – is now being overcome. And the compatibility problems between early MAP versions may also be about to disappear. At first GM said that the upgrades from MAP 1.0 to 2.0 to 2.1 and 2.2 must be compatible one with another, allowing users an easy transition or migration path from each version to its successor. But GM also said that there would not necessarily be an upgrade or migration path available from MAP 2.X to the definitive MAP specification, MAP 3.0. Now this picture is changing. Specifications are being developed for hardware, software or both which will enable users to convert MAP 2.1 equipment to MAP 3.0 (see 'Migration', section 6.1.2).

One other 'problem' is more apparent than real. As this book explains, further MAP variants will be available which use either part or all of the OSI model. But these variants or subsets of MAP, which include 'carrier-band' MAP, the 'enhanced performance architecture' (EPA) and the field bus, will be compatible with the main MAP specification: as will also be explained in this book, these variants are designed to perform special tasks in connection with the full MAP system.

The cost of MAP systems is coming down. Some carrier band MAP systems became available during 1986. These versions of MAP may be the choice of many small companies who want to invest in MAP systems as a way of making it easier to add future capacity. Costs at the end of 1985 were anything from $2,000 to $10,000 per connection. A year later they were about $1,000 and due to fall to between $200 and $500 or below when the very large-scale integration (VLSI) chip versions of MAP appear in quantity.

The completion date for the jigsaw of standards which will make up MAP is 1988, by which time most suppliers will be offering plug-compatible, MAP standard, VLSI-based communications products. When that happens, the era of computer-integrated manufacturing can be said truly to have begun.

2
CURRENT MANUFACTURING TECHNIQUES

At the end of 1985 International Data Corporation (IDC) put the total worldwide industrial automation market at $11.3 billion, growing to $39 billion in 1990. If these figures are accurate, this growth is immense. But more interesting even than the growth of the market is the way the figures are broken down. This shows very clearly that, while robots, for example, are stuck at two per cent of the market in 1985 and 1990, manufacturing information systems – in other words the systems which move design, production and other data in and around the factory and its related offices – will grow from under a third to nearly 40 per cent of this much greater total market, a higher growth rate than any other part of the market.

IDC is not the only market research company saying this. Consider, for example, a report produced by Frost and Sullivan in the autumn of 1986 which put the growth of the market for factory local-area networks (LANs) at 35 per cent. At this rate the market will grow from $51 million in 1984 to $431 million in 1989, when the US alone is expected to have 41,000 LANs connecting up one million devices in its factories: in 1986 the number of LANs in the USA was just 1,600.

The figures from another US research company, Venture Development Corporation (VDC), are more conservative. VDC has said the 1984 network market was already $135.2 million, so that in rising to $473 million in 1990, the growth rate is lower but, at 23 per cent, still respectable.

Market reports are not wholly reliable and IDC's figures, for example, have been overtaken by intervening cutbacks at General Motors. But what these analysts are saying is entirely logical, because their findings reflect a coincidence of ends and means. The end is computer integrated manufacturing (CIM) and one of the means to achieving it is the Manufacturing Automation Protocol (MAP). The market research confirms this. VDC's conservative figures, for example, show the MAP proportion of the total factory communications market rising from 0.7 per cent in 1984 to 78 per cent of the market in 1990.

There are good reasons for this. In the very early 1980s most of those in manufacturing were convinced, or advised, that the way forward was a simple reduction in labour costs by replacing men with machines which did the same job as the men they replaced. This meant the straightforward adoption of robotics and robot-fed flexible manufacturing systems, plus a computer aided design (CAD) screen or two and a separate materials-requirements or manufacturing resource planning system (MRP1 or MRP2) to schedule the whole thing. Naturally the robot manufacturers were keen to promote this kind of thinking and they were helped by consultants who thought the same. Anyone who considered linking up these separate 'islands of automation' into a computer integrated manufacturing (CIM) system gave CIM a moment's thought and then dismissed it as a flight of fantasy.

But, for once, it may be that the conservatism of the accountancy profession has been working with manufacturers instead of against them. Many such islands of automation as those just mentioned were installed but, perhaps because of conservative approaches to accounting, nothing like the numbers that the hardware suppliers had been hoping for. The ones that were installed produced mixed results. Now the use of technology to 'reduce costs' by throwing technology at existing problems is regarded as automating the problems, not solving or eliminating them. As the saying goes, automating a bad process makes it run faster. It does not make it run more efficiently.

So, since about 1984, manufacturing industry and its suppliers appear to have decided that the integration of the separate functions in a manufacturing operation is not only desirable but, given suitable methods of moving data around the operation, possible. MAP has emerged as the main candidate for a 'suitable method' and, as such, is one of the reasons the manufacturers and suppliers have changed their thinking.

It is difficult to overstate the importance of the change we are about to describe. It transforms conventional approaches to manufacturing technology. CIM can be defined in various ways but, if we think of CIM as the linking of hitherto unrelated activities in a manufacturing company, the CIM approach is much more mature than the 'islands' approach since, correctly applied, CIM embraces the whole business of which manufacturing is a part. The technology-led approach that preceded CIM assumed that a business could be made successful by listing the areas where there were production problems or bottlenecks in order of the urgency with which they must be solved and applying technological solutions to each in turn, in isolation from each other and from the business as a whole. CIM moves away from this idea. This does not mean banishing technology completely. It means assessing technology's value in business terms rather than for its own sake. This means that the manufacturers who take heed of the changes are on the brink of a phase in industrial development potentially as important and exciting as the invention of the steam engine or Henry Ford's production line.

This is partly because, even if, in the beginning, microprocessor based technology was not all that well applied in industry, it is at last becoming familiar in today's factories. The modern machine tool is controlled by a computer based numerical controller. Conveyors, ovens, welding machines and high-volume so-called 'fixed' or 'hard' automation systems are controlled by various micro-based programmable controllers. Robots have their own micro-based multi-axis controllers.

The first compelling argument for effective communications systems is the development of off-line programming systems. These micro-based shop floor devices are no different from other computers, they have to be programmed. One way to illustrate some of the problems and opportunities the micro presents in manufacturing is to examine the process of programming a conventional machine tool. The usual way of doing it is to run a punched paper tape through a tape-reader on the machine. The tape reader converts the information in the holes into a form it can store in its memory. Punched tape has a number of disadvantages. If used repeatedly the tape itself may become damaged and unreliable. Mylar can be used instead of paper but is expensive, especially if, as often happens, mistakes are made when punching the tape. The tape reader on the machine is usually unreliable. Paper-tape programs often do not allow changes in speeds and feeds to be made during cutting so these must be set for worst-case conditions. This is inefficient. The controller is hard-wired and so must be completely replaced to provide the machine it controls with any new features that come along.

Computer numerical control (CNC) dispenses with all these difficulties and provides management information and machine diagnostics as well. The tape is used only once, to enter the program, and thereafter the program is stored in the machine's memory. CNC allows the tape to be checked on the machine and optimized for different feeds, speeds and so on. New control options can be introduced merely by updating software. And users can write their own sub-routines as 'macros' for storage in the CNC memory. Even more important, a CNC machine lends itself to communicating with a plant-wide supervisory system.

This is accomplished through direct numerical control (DNC). The tape reader is left off each machine and the part program is sent to the machine tool from a supervisory computer. The main point of DNC is that it is used to control a number of machines, though it has been applied in shops where there is only one machine so that paper tape can be eliminated. All the programs for all the machines are stored on the mass storage system linked to the supervisory computer. The supervisory computer calls the programs off a disc as the machines need them. The machines may be multi-tasking; a program can be sent to a machine while it is still machining a part according to a previous program.

Just as important, the machine can send data back to the supervisory computer about which part it is working on, what information it has about the number of parts needed, when they are needed and so on. It can tell the supervisory computer whether it is machining, loading, unloading, waiting for parts, waiting for tools, jigs or fixtures or has broken down. All this information can be collected and presented as a histogram showing how much of its time in use has been spent over a given period in these and other various conditions. The analysis can be made over an hour, a shift, a working week or longer. The machine can also send signals about how long ago it was repaired or maintained and send an alarm if it is due for another maintenance.

If there are many machines in a factory the supervision may be delegated to a number of smaller computers all linked into a larger system. This means a communications system is needed which exchanges data between the main computer and the smaller supervisory computers. Links are also needed between each supervisory computer and each of the machines it supervises. Until now, these links would have been proprietary links using data exchange protocols only the computer supplier's machines could understand. The systems builder, usually not the same as the supplier of the computers, would have to write software interfaces to enable the supervisory computer to exchange information with the controller on each machine. This task is inevitably complicated by the variety of machines that may be in use in any one manufacturing operation. No single machine tool vendor can cope with the entire range of machines that any moderately large operation needs to use, from pipe-bending to punching sheet metal to turning and other operations. For example British Aerospace of the UK reports that its machines varied from modern CNC to hard-wired NC to very old machines controlled by analogue circuits. Each controller has had to be modified so that the signals from the outside world are seen by the controller in the format and protocol the controller understands and can operate on. In the case of the machine tools there were 50 of them with 10 different types of controller.

The source of the programs which will run the machine-tools is the computer-aided design (CAD) system on which the parts the machines will make are designed. The CAD system is used to create the design of the finished part. The programmer turns it into the sequence of step-by-step operations the machine-tool uses to make the part and post-processes it for the particular machine tool being used. Often the design and programming functions are carried out in separate departments so this implies effective links between the CAD and the programmers' terminal.

The CAD system can support other functions. It can supply a program for a probe on the machine tool to check some measurements after a part has been made. The data can be fed back into the machine tool controller to compensate for tool wear. This makes the machine much more accurate and consistent – it means the last part in a batch is just as accurately machined as the first. Or the CAD system can supply a program for a co-ordinate measuring machine (CMM). This is a numerically controlled arm, usually mounted on a gantry, which can give one part in ten or a hundred a thorough dimensional check. The data from the check can be fed back up to the supervisory computer to provide trend analysis. The increasing use of statistical process control in factories means a CMM can stop a whole production line if a series of parts looks likely to go out of tolerance. This eliminates a potential problem before any parts have to be scrapped.

Quality control data can also be fed back to the supervisory computer by human inspectors. These may carry out detailed examinations of, say, a car and feed information into a handheld terminal or – even better because it leaves their hands free – by spoken commands into a voice recognition device. Intel demonstrated these voice-recognition techniques at Autofact 85 in Detroit and they are used in many US car factories and at Jaguar in the UK.

Even if machine tools are the main programmable items in a factory there are also likely to be programmable controllers (PLCs) and there may be robots too. The PLCs which control conveyors or carry out other tasks are usually easily programmed provided the user understands their programming language, usually an analogy of the way the old relay control panels were built before the microprocessor was invented. If, as often happens, a number of PLCs have to be linked to carry out a particular sequence of tasks along a production line, the links between them are, again, usually unique to the PLC supplier. Examples include the Texas Instruments Tiway 1 and Allen-Bradley's Data Highway. Each time a new PLC sub-system goes in, a new network is installed which does not interface with other makers' systems unless you make it do so.

Robot programs are more complex. Robots are yielding to off-line programming using simulation programs of various sorts which can home in on tiny robot movements and reproduce them accurately in the program. But these programs usually assume 'perfect' robot arms which do not flex or judder to a halt even when carrying the highest payloads. Nevertheless, robot simulation is an obvious case for good communications between robot and remote programming system. A robot is unlikely to be used by itself so its controller has to be interfaced with the control system used for the cell or system the robot is working in.

Similar special problems have to be solved to integrate automatic guided vehicles (AGVs), vision systems and other plant items. Indeed, the intelligence the microprocessor provides is powerful mainly because it is now cheap enough to incorporate on almost any machine or plant floor item. But if this intelligence were used in combination with the intelligence on other plant items and systems, this intelligence would become still more powerful. For example, when relays were used to control equipment, the controller was given a fixed routine or set of instructions and went ahead on those instructions whatever happened. Now electronic logic and computing power is so cheap that machines can 'react' to changes. If a machine fails they can see if there is a way the operation can continue without it.

Indeed the successfully automated factory will not even begin to make a part or product until it has checked the inventory to make sure that all the materials or parts needed to make it are ready. It will also check all the machines needed and, if one of them is out of order or unavailable because it is making a part, it will not start the process for the requested part.

Cheap intelligence makes it possible for machines to cope with a variety of different products. So where once a machine or set of machines could only make one thing in large volumes, it is now possible for the same set of machines to make a variety of parts or components.

This capability changes entirely the economics of production. Many now involved in manufacturing insist that traditional accounting methods, particularly those which determine a time within which the investment in an item of plant equipment must pay for itself, are completely unsuited to recognizing the benefits new production technology can bring. The reason is that a system which can make one part as easily as any other, within the limits described below, can make parts in random order on an instruction from, say, a bar code reader or a voice recognition device. So it can make one component, then go on to another and then a third, and may not make another of the first type of component for days or weeks.

This means that the economic batch quantity (EBQ) the minimum number of a part which a factory can make economically, goes down to its theoretically ideal size of one. This flexibility should not be taken too far. It is pointless to aim, for example, at a production system which can make pram wheels one minute and fuel pumps the next. Rather the equipment should be organized to make a family of similar parts. The car industry is a good example of this. A car factory now supplies so many varieties of colour, external and internal trim and optional extras that each car could be considered as a batch of one. Even the car industry is now rationalising, however, to reduce the number of possible options without reducing the perceived choice available to the customer. An example of much smaller scale flexibility is IBM's typewriter factory in Lexington, Kentucky, USA. This makes ten variants of the same product. The basic product is altered to turn it into a typewriter or a computer printer, depending on the order the shop floor receives.

But intelligence has to be used sensibly at corporate level too. For example, the barriers between design and production departments are there as much for social as for technological reasons but technology can demonstrate how costly these barriers are. The savings that can be made by improving communications between these departments are hidden but can be immense. Design departments are not yet used to designing products so that the production department can make them easily or make them with automated equipment. A combined design and production department, or two that can interact as one, could eliminate expensive design oversights. This can work whether the factory is automated or not. Ford has found that some products designed for automated assembly were so easy to put together that it was cheaper to assemble them by hand! Design departments are also wont to make a series of apparently minor changes to designs because they have just thought of a better idea. Often such changes cost a great deal in lost production and add the cost of new fixtures, new parts which may be in stock, old parts which are in stock but no longer needed and so on. Managements are telling design departments the design stays as it is unless there is a very good reason for updating it, like a safety considerations. Using conventional paper-based systems, when designs are allowed to be changed, there is a delay while the knowledge of the changed reaches everybody affected by it. It is common to hear designers speak of design changes taking two weeks to reach somebody whose previous two weeks' work is therefore totally wasted as a result of the change. If the change takes such a time to reach a supplier who is making a part under subcontract, the supplier will expect to be paid even for the wasted work.

One approach to making all this possible is to develop a database which holds a complete model of every product the company makes. The designer does all the work on a CAD computer screen and, when the design is complete, sends the finished design to a database for storage. Thereafter, every job function within the operation extracts information from that model. The database provides information for parts lists, bills of materials, the sequence of production steps needed to make parts, and supervisory information for the shop floor functions which will make the finished product.

This approach offers great savings in time and materials. It means manufacturers can build up complex products, like cars, much quicker because several different parts of the car can be designed and made at once instead of sequentially, i.e., parallel build. But it also implies that the whole production process and the ancillary functions round it have been carefully rethought from top to bottom. And it implies that the company has to work out communications systems that will make it possible.

The communications systems are in two parts. There is the human communications path: this implies that barriers between departments – marketing, design, production, sales, accounts, have to go. Almost as difficult is to decide what communications system will link the various computer-based systems the company already has. This applies in particular to the corporate database on which the successful management of a company may depend. The leap of imagination needed is to see the modern manufacturing operation as a series of interlocking functions. Modern manufacturing is not about materials requirements planning (MRP), or flexible manufacturing cells or computer aided design systems or office automation or computerized forecasting or the latest computer based accounting system. All these are useful management tools but all too often they are isolated systems. And their very isolation renders them less effective than if they were treated as components in a manufacturing operation which is itself treated as a complete organism.

Existing systems usually produce a printout which is taken to another department in a company for processing there. This means a bill of materials, for example, may be rekeyboarded many times and the chances of an error occurring are very high. Some of the mistakes caused can be expensive. A typing slip in a materials order can mean unnecessary material costs incurred, high inventory and consequent high interest charges. Not enough material ordered, on the other hand, may result in delayed, and possibly cancelled, contracts.

Design changes have to be controlled. The changes that are permitted must be communicated instantly to those affected by them. Computers can produce lists of those affected by a change in the design of a particular part. With effective communications, the computer can disseminate the change throughout the organization.

Reducing the time taken to achieve such changes can reduce costs and can make the difference between, on the one hand, gaining market advantage by putting products on sale months before the competition and, on the other, being the competitor that has to do the catching up.

All this means that effective, competitive production is no longer just about the application of computers in industry, important though this is, but about the integration of these effective but separate computer-based elements. In turn, this means that communications is the framework on which the profitable manufacturing company must build. As luck would have it, communications is the first discipline in which the users hold the whip hand over the vendors.

3
THE NETWORK
The Physical Transmission Medium

3.1 Basic Communications

Those seeking to explain the Manufacturing Automation Protocol (MAP) and its office equivalent, the Technical and Office Protocols (TOP), have struggled to find comparisons that will make them easier to understand. A good analogy is that with the ring main or domestic power circuit. Everyone in the more developed countries assumes access to domestic electricity, and would find it difficult to carry on their lives without an electric socket. They would also assume that, at least within a country, most sockets have a standard shape and size, no matter what has to be plugged into them. And they would assume that the same socket could power a lamp, an electric typewriter, a washing machine or a drill. There is a limit to the analogy, but it does correspond roughly with General Motors' original conception of MAP. MAP began as a central communications core or 'bus' into which every piece of equipment for the factory floor must plug. If the equipment did not plug into the MAP backbone GM originally conceived, then GM would not buy that equipment.

However, this idea could not have held for long. It meant that GM would have to pay for a MAP-capable plug and socket for every programmable system used in GM – at $1,000 for each MAP 'interface', GM's 1990 bill would be $200 million for MAP interfaces alone. Sooner or later GM would face a choice between sticking to this MAP 'backbone' or compromising. As will be explained, GM compromised.

A better analogy might be that with a railway carrying tools, or components or sub-assemblies which need final processing or assembly in a factory. A train collects goods at any station on the network and carries them to any other station on the network according to labels attached to the goods. A MAP bus does a similar job to the railway. Railways cannot function unless all their rolling stock uses the same track-widths and loads and unloads to the same platform height. This is the kind of standardization that MAP and TOP are about. Goods arrive at a station by all sorts of methods – on foot, by bus, by taxi or by car – and when they get to the station where they leave the railway they may be carried on to their final destination by a similar variety of methods. At the station where the goods are taken off there might even be a light railway which links it to the factory. The light railway may or may not use similar rolling stock to that used by the main railway network. But, whatever happens away from the main railway, and whatever the goods it carries, the main railway always uses the same track and standard rolling stock.

Once the goods enter the factory any resemblance to the main railway diminishes even further. Any number of methods may be available to move around the raw materials collected from the station and the same systems also move finished parts around the factory for eventual delivery to customers, again using the main railway. The factory may use rail-guided vehicles, conveyors, automatically guided vehicles, overhead cranes, forklift trucks or a mixture of any of these. The raw materials coming in from the railway and the goods leaving by it remain the same, though labels on them may be added or removed at various stages of the journey.

The MAP broadband cable is the equivalent of a main railway which is carrying, say, tools around the country. The MAP railway carries packets of information, frames of data, without which an action at the final destination of the tools cannot be carried out. Other versions of MAP, which will be explained elsewhere, and various proprietary, non-MAP networks correspond to the communications systems which link, first, the MAP main railway with the factory and, second, the conveying systems within the factory – the AGVs, conveyors and so on. The stations where the 'goods' are transferred to the vehicles or light railway taking them to the factory are the 'gateways' to the main railway. This and the other types of connection in a MAP and TOP communications network are described later on.

One of the most important things to note about MAP is that it does not influence what is carried on the railway, it only carries it. The railway's function is to make sure that the goods, whatever they are, arrive at the destination in the same condition as they started. If, when the tools arrive at the factory, they do not fit the machine they are intended to fit, it is not the fault of the railway. In just the same way, MAP does not determine the fitness or otherwise of the data entrusted to it. It is therefore not valid to expect MAP to sort out incompatibilities between the different systems which are connected on a MAP or TOP network. For example, some potential MAP or TOP users expect MAP to sort out the incompatibilities between different CAD systems, doing away with the need for the neutral format provided by the initial graphics exchange specification (IGES). Neither MAP nor TOP can do this, any more than a telephone can translate the speech signals of a call in French or German into English.

Bridging databases is another task MAP and TOP cannot solve alone. An increasingly common task is for one database user to try to gain access to data held in another database, or even a third or fourth database. Neither MAP nor TOP can make easier the business of gaining access to these databases, each of which probably uses a different operating system, needs different passwords and has a completely different method of storing and retrieving data. This problem of incompatible databases is being tackled by the development of 'distributed systems'.

So what is the point of MAP and TOP? They provide the local area network (LAN) over which compatible systems can talk, *regardless of who supplied the systems*. LANs allow the sharing of resources without the need to plug and unplug terminals, printers, disc drives and so on. Devices can easily be added to or removed and, once a device is tapped on to the LAN, it is available to all the other devices on the same LAN able to make use of it. LANs can be connected to databases, so that information can easily be fed from one database to all the devices that need it. What is more, the information can be updated, fed back to the database and, thereafter, anyone who wants access to that information will immediately receive the information in its updated form. In a design application, for example, there is no need to notify all those holding drawings of a design update. LANs can also be connected to other LANs in the same company, in the next building, across a city (in a metropolitan area network or MAN) or in a different country by satellite (a wide area network or WAN). All these networks may also provide their own internal electronic mail services to exchange memos, provide diary information and the like. LANs may be used to exchange voice, facsimiles, video and text information as well as data. These different types of information may be carried together on separate channels or, by multiplexing, on a single channel system.

3.2 Communications options

Before describing MAP and TOP in more detail it is worth examining some of the choices that were open to General Motors when it set about describing a general purpose factory LAN. A data communications system can be categorized by:

the medium, or type of cable, it uses,
the shape, or topology, of the network,
the method of access available to each node on the network.

There are three main types of cable media: twisted pairs of wires, as in a telephone cable; coaxial cable; fibre-optic cable.

3.2.1 Twisted pair
The three media types vary in performance and flexibility. The most common transmission medium, as it is used to carry telephone traffic, is twisted pair wire, i.e. two insulated wires twisted round each other. Usually multiple pairs are bundled into one shielded cable.

The twisted pair has the advantage that every organization with a telephone system is already wired up. Most telephone systems have two, three or more pairs of wires going to each telephone. But the phone itself still only one pair and the others are available for other uses. The sound waveform made by a voice signal, for example, can be transmitted as a series of binary numbers sent in rapid succession. 'Digitizing' voice signals in this way makes them much easier to recover intact even if there is a lot of interference in the system. And short spoken messages are easy to store in digitized form. The telephone wires can send and receive voice signals broken down into binary digits – 'bits', a logical 0 or 1 – at a rate of 64,000 bit/s.

Computer files and other data can also be sent on twisted pair, though at a slightly lower real rate because extra data has to be added to provide signalling information. The data rate is 56,000 bits per second.

The rate at which communications systems may transmit is the source of much confusion. One reason is that the available speed falls as the length of the network increases. But the main source of confusion is that there are ways of sending data which are actually faster than the given maximum bit-rate. The reason is that, even though a signal may contain only one bit, that bit might contain enough information to be decoded into two, four, eight, 16, 32 or even 64 different states. This is because the voltage of the bit may have a number of different values, and each of these values may also contain 'phase information', which effectively means there are a number of subtle timing variants which give the pulse further different values. All these values can be decoded with the right receiving equipment at the other end. This is why systems based on twisted pairs can be said to operate at 128,000 or 256,000 'bits a second' or even higher in special circumstances, but in such cases the 'bits a second' usually turn out to mean 'bauds', not the same thing as bits per second. As explained above, the equivalent of several bits per second can be crammed into one baud. Another consideration in trading off speed against distance is that much of the theoretical cable length available is soon used up in a building which has high ceilings or covers several floors.

One limitation to twisted pair LANs concerns the second communications system characteristic: its topology or shape. A network based on the telephone system is star-shaped, with every phone going back to the exchange at the centre. The data may not be routed through the same switching system as the telephone calls, but it goes back to a separate data switch at the same place as the telephone switchboard or a data switch built into the switchboard. This means that the failure of a device on the network does not affect any of the others but it also means that, if the data switch collapses, so does the network (section 3.3.2). Access to such a star-connected system is governed entirely by the throughput of the data switch. Twisted-pair communications systems are also 'hard-wired': they have a limited capacity for expansion and wires need to be rerouted and reconnected if a 'node' – any device connected to the network – is moved.

Network management in such a system can be complex. If a lot of devices are linked into the same network some devices may have to wait for long periods if the node they want to talk to is already talking to some other device. If queues are to be avoided, the network has to be speeded up to get through the data that has to be shifted. This means using higher frequencies and, again, once the frequency reaches a certain level, the distance over which the network may transmit is reduced. This is because higher frequency signals migrate to the outside edge of a conductor – the 'skin effect' – increasing the effective resistance of the wire.

3.2.2 Common connecting methods
In most cases the spare wires in telephone lines or PBX-type data switches are just not convenient to use. They are either in the wrong place or the user wants to be able plug devices together and unplug others. In such circumstances probably the first system to be considered will be RS232, a communications standard in common use since its adoption in 1969. RS232, which is really a more sophisticated variant of the twisted pair, was developed jointly by a committee of the US Electronic Industries Association (EIA) and Europe's telecommunications standards body, the CCITT, part of the International Telecomunications Union agency of the United Nations.

RS232 was designed to send data from a terminal to a modulator-demodulator (modem). The modem then converts the signals so that they can use the public switched telephone network to talk to another modem, and that was connected to the receiving terminal by another RS232 interface. Now it is used for everything from connecting terminals to computers and sending instruction to programmable controllers. Every personal computer has an RS232 port (plug-and-socket) to connect it to the things that make it useful, such as a printer or another computer.

Although an RS232 port has a 25-pin plug and socket, it is really a two wire, single channel system with extra wires for control functions which allow the interface to set up a data transmission sequence between one device with an RS232 port and the similarly equipped device it is calling. The fact that it is a 'standard' means the user should be able to plug in any printer whose features he or she happens to like and expect it to work with their computer. The alternative would be to restrict users to the make of printer their computer happened to fit. The standard determines the size and number of pins on the plug and its socket, what the voltages (within wide limits) of the electrical signal passing between the printer and the computer should be, and how the link can be connected to other systems. RS232 is in fact a cross-fertilization of CCITT, EIA and International Standards Organization (ISO) recommendations created at various times. Some pin-assignments in the RS232 connector, for example, were allowed to be specified by particular countries. Because of this and because the standard does not, even then, specify everything the communications engineer needs to know to establish an RS232 link, RS232 links can vary. The standard does not lay down the method for controlling the flow of information across an interface. For example, how does a receiver tell the transmitting terminal to stop while it clears its buffer of the data already received? RS232 specifies four signal types: data signals, control signals, timing signals and automatic calling signals. But some circuits need all of these and others use only some of them: the only really essential RS232 connections are 'Ground' (pin seven), 'Transmit data' (pin two) and 'Receive data' (pin three). The voltages can vary between plus or minus 3 volts to plus or minus 25 volts. One communications engineer's remark that he had never come across two RS232 implementations that were the same is probably exaggerating the problem, but not much.

RS232 usually sends its information at 9,600 bits every second but the standard specifies signal rates as high as 20,000 bits per second. Again, however, the signal begins to weaken if it is used at a high rate over long distances – it is designed for terminals no further away than 100 metres, though in practice it can be used at twice this distance.

RS232 is an unbalanced system, which means that the transmitting and receiving devices use an earth or ground return. This makes it easy to set up but, if the so-called 'common' earth is at different values at either end, this results in what is called 'common-mode interference' because of the voltage difference between the two earths. This could be overcome by increasing the signal voltages, in some cases as high as plus or minus 80 volts. Such large voltage swings made RS232 impractical for all but the slowest signal exchanges.

Attempts have been made to replace RS232 with a signalling system that performs better. In 1977 the EIA came up with RS449. This standard defines signal functions and connector details but leaves the description of the circuit to two other standards, RS422 and RS423. In RS422 the logical one is a voltage between two and six volts and the logical zero is one-fifth of a volt or less.

RS422 can use either twisted pairs or coaxial cable. It is a 'balanced' system, which means it does not use an earth return and can be terminated in such a way as to avoid signal reflections caused by mismatches (see section 3.3.2 on collisions in Ethernet systems). This means RS422 can operate at one thousand times the rate of RS232 over distances up to one kilometre or more. It also uses zero and (about) five volts as the electrical equivalents of the logical 1 and logical 0 machine language that digital devices use; RS232 uses anything from minus three to minus 25 volts or more for a logical 1 and plus three to plus 25 for logical 0. This means the RS422's power supply is simpler and, therefore, cheaper.

Another important feature of RS422 is that it allows the connection of one driver station to up to ten receiver stations: RS232 allows only one driver and one receiver. Although RS422 is being developed further to allow it to operate more like a bus – which means devices can be added to it and taken away as required – it is still not flexible enough for most communications purposes. Its driver chips have been more expensive than the RS232 drivers and, though RS422 is used in special circumstances, its progress in replacing RS232 has been slow.

3.2.3 Other twisted-pair systems
There are one or two other communications standards of a similar type which should be mentioned. The 'current loop', in which the presence of a current of 20 milliamps represents a logical one and its absence represents a logical zero, was originally designed to drive electromechanical parts in teleprinters. Current loop systems can provide high noise immunity but 20 milliamps is quite a high current. The link is terminated by a voltage sensing device. This means that resistance

along the circuit must be kept low and line lengths quite short if the signal is to be large enough to be recognized unambiguously at the other end. Current loop systems often use optocouplers at either end of the link and the performance of optocouplers deteriorates as they get older. The field bus (see section 4.8.5) is the intended replacement for current loop.

All the twisted-pair systems are 'serial': each eight-bit long data 'word' has to be passed to the printer one letter at a time along a go-and-return pair of wires. Two directional communication requires the provision of a third wire. A faster but more expensive alternative is a 'parallel' connection, which gives each of the eight bits in a digital word its own wire, so all eight bits arrive at their destination at once. IEEE488 is such a system. Normally its use is confined to instrumentation systems because switching pairs of wires is much less complex and expensive than switching among groups of eight parallel wires. Parallel cabling is, in any case, inherently much more expensive than using twisted pairs or coaxial cable. Techniques such as time division multiplexing (see section 3.3.4) allow the distribution of what seem to be parallel signals along a serial coaxial cable. Note too that, in a baseband system, two-directional communication normally means the provision of two sets of cables, whether they are serial or parallel.

Although, as will be explained, there are MAP-related communications systems which offer alternatives to these commonly-used communications standards, RS232, RS422 and current loop will continue to be used where appropriate. Most programmable controllers, for example, now have one or more RS232 ports or offer a choice of RS232 and RS422 and this is likely to continue to apply. Controllers and other devices linked with such systems may be connected to a MAP network through a 'gateway' (see section 4.9).

3.3 Distributed LANs

Strictly speaking, even the lowliest twisted pair system is a local area network (LAN), usually star-connected (see Figure 3.1). The topology or shape of a network gives it certain inherent access characteristics. In a star network access is through a switch of some kind, and a means has to be provided of dialling through to the right destination. Buses and rings (Figures 3.2 and 3.3), however, can be 'fully connected', which means that every node is connected to every other node. One of the chief characteristics of bus and ring LANs is that they have provision for addressing and can deal much more easily with data exchange between large numbers of devices. Each extra device has a new address, every device on the LAN reads the address in each data packet coming down the LAN. If the address for the data coincides with its own address it takes the data; if it does not, the device either copies it from the LAN or takes it in and passes the data back on to the LAN. All messages go to every node, but addresses added to the message make sure that only the node for which it is intended actually uses the message.

This is why the usual definition of a LAN applies to systems which use 'bus' or 'ring' topologies to provide services beyond the capacity of

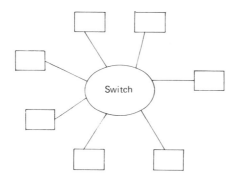

Figure 3.1 Star-connected local area network

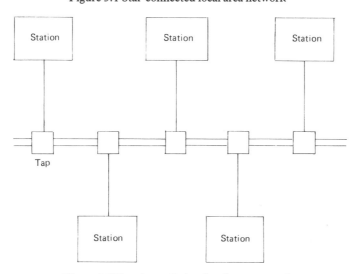

Figure 3.2 Topology of a bus local area network

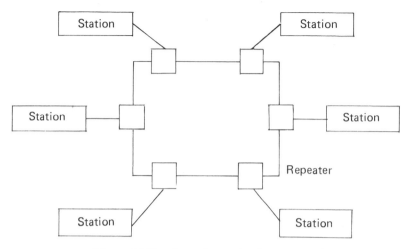
Figure 3.3 Topology of a ring local area network

star-connected system. For the sake of convenience, this is how 'LAN' will be defined from here on. A bus is a length of cable with nodes or gateways connected along its length. A ring is a bus with the two ends joined together. These LANs usually operate at higher transmission rates (10 million bits a second or more) than twisted pair networks.

Ring networks are unidirectional but, unless the ring is broken, all the nodes receive all the messages eventually. All the nodes in a ring must be able to receive some messages and pass others on, so they take messages in and regenerate them before passing them on. In other words the node has a built-in repeater. In a bus, the messages must pass in both directions if all nodes are to see them but each node merely samples the messages, leaving untouched those not addressed to it. This means bus networks use rather simpler interfaces to the network.

Transmission rates of 10 million bits a second or more are usually achieved using coaxial cable. Coax consists of a conducting wire inside a second, hollow conductor and separated from the hollow conductor by an insulator (Figure 3.4). The hollow conductor is 'earthed' and shields the signals on the central conductor from interference. The coax's construction also makes it less susceptible to the 'skin effect'. These two features contribute to the coax's improved performance.

The coaxial cable can be subdivided into baseband and broadband types. Baseband means that the data is fed directly on to the network and can be picked up and understood in its original form without complex demodulating equipment. Twisted pairs of wires, as used in telephone systems, are always baseband, whereas coaxial cable can be either baseband or broadband.

The Network – the Physical Transmission Medium 33

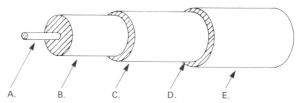

Figure 3.4 Coaxial trunk cable

A. Centre Conductor: Centre most feature of coaxial cable, it consists of solid copper or copper clad aluminium wire.
B. Dielectric: Electrical insulation utilized to maintain position of the centre conductor. It is composed of foamed polyethylene. This insulator/positioner may also be evenly spaced polyethylene discs.
C. Outer Conductor: is constructed of an aluminium tube. Cable size (412, 500, 750 & 1000) is derived from its outside diameter.
D. Flooding Compound: (Optional): A viscous substance placed between the outer conductor C and the jacket E to maintain a protective seal should the jacket E contain or develop any cuts or openings.
E. Jacket: (Optional): a black polyethylene coating over the aluminium outer conductor to provide a weather tight seal.

There are two types of coax in common use: 50 Ohm and 75 Ohm. The 50 Ohm cable is used for digital transmission only, whereas the 75 Ohm is used for both digital and analogue transmission. The 50 Ohm will support speeds of up to 10 million bits per second (10Mbit/s), whereas the 75 Ohm will go up to 50Mbit/s. Typical cable for MAP application will be 75 Ohm semi-rigid RG-6 or RG-11 coax running 5Mbit/s or 10Mbit/s data rates.

Bus and tree (Figs. 3.2 and 3.5) shaped networks are commonly based on coaxial cable, which is slightly more expensive than twisted pair and the distance covered depends on the transmission technique.

3.3.1 Fibre optics

A fibre optic cable consists of a very thin, very flexible glass or plastic cylinder down which light can be transmitted by reflection from the internal walls of the cylinder. Fibre optic cable offers very much greater speeds and distances than coaxial cable and its immunity to electrical interference makes it invaluable for industrial and military uses. Fibre optic's major problem is no longer its cost but the proliferation of different standards for the cable itself and the light source which transmits the signal.

There are two types of light source, light-emitting diode (LED), and the injection laser diode (ILD), and broadly two types of cable, single-mode and multi-mode. Each cable type can support either LED or ILD sources. LEDs are cheaper but offer much lower power than ILDs and so are suitable only for short distances. ILDs are required for

34 MAP and TOP: Advanced Manufacturing Communications

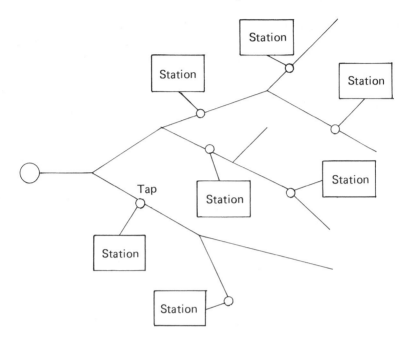

Figure 3.5 Topology of tree local area network

higher speeds (greater than 1000Mbits/s can be achieved) and longer distances. Multi-mode has lower bandwidth because the internal path-length for the light can vary, so multi-mode is more suitable for short distances and single-mode for long distances.

It is not possible to mix the different fibres as the connecting plugs are different. Some attempt at standardization has been made in the USA under the American National Standards Institute (ANSI) who have defined a 100Mbit/s LAN called FDDI, the fibre distributed data interface, which uses multi-mode fibre. Another branch of ANSI, however, the Exchange Carriers Standards Association (ECSA), has defined a Synchronous Optical Network (SONET) which uses single-mode fibre.

In terms of baseband and broadband definitions fibre optic systems cause some confusion because, although fibre optic transmission does use a carrier – a light beam – it is usually referred to as a baseband system because there is only one carrier – the light beam – and it has a constant frequency. There is, however, no reason at all why in the future different light-frequencies should not be used together on one fibre-optic cable.

Fibre optics seems to offer a lot of attractions but Ferranti of the UK, which has put in a number of broadband systems, does not feel that fibre optics yet offers any advantages over broadband. The cost of a connection to broadband is cheaper than fibre, the company says. Ferranti cites the difficulty and cost of splicing two lengths of fibre cable together: broadband splicing is much easier and so is the connection of nodes to the cable. Neither does Ferranti accept that fibre even offers any noise immunity advantage: one Ferranti engineer says: 'The noise immunity of broadband is such that we never need to put it into conduit.' Fibre also goes opaque when subjected to high radiation – cameras in nuclear stations need special lens glass to counteract this.

Even those in the communications supply industry who agree that fibre optics – usually referred to as 'fibre' – is the eventual chosen medium also point out its current drawbacks. One is that fibre technology is still 'fluid'. For example, the topologies of fibre networks can vary just as those of copper based networks can. The two topologies most often considered for fibre are stars and rings: fibre optics is unidirectional so its use in buses is limited. In a star network (Figure 3.1) there is a central hub from which the fibre optic connections radiate. This hub could be 'passive', a lump of glass to which the fibre optic spokes are attached. Or it could be 'active'. An active hub converts all the fibre optic signals from light pulses to electrical digital form, processes and monitors them, then reconverts them to light pulses and passes them down the spokes.

The passive hubs supplied by companies such as Codenoll, for example, need no power supply and are almost totally reliable. The active hub allows better network control but if the hub fails the whole network fails. And consider what happens when a node on a fibre ring fails. Using ordinary copper links a node can easily be bypassed. Using fibre optics, a failed node usually means that the whole ring stops working. The only way to arrange bypasses is to take a monitoring line from each node to a central network manager which can detect if a node has stopped. The network manager then passes the signals round the failed node. But what form should these monitoring lines take? Should they be fibre or copper – either is possible.

And, whether the topology is ring or star, at what speed should it work? All these things have yet to be agreed before a fibre optic MAP standard can be published.

3.3.2 Ethernet
The most common type of LAN used is, in fact, normally a 10Mbit/s baseband system, Ethernet. Ethernet is a 'bus' developed by Xerox with Intel and DEC, then taken up by the US Institute of Electrical and

Electronic Engineers to become the 802.3 standard. Ethernet uses 'multiple access', which means that every node along the Ethernet bus may transmit to the network at any time. Each node has an address which distinguishes it from any of the other nodes on the network. When a node is transmitting it broadcasts its message equally to every other node on the network. The message begins with a preamble which sets up the timing of the receiving node followed by the address where the message is to go. If the message is for all stations the address is all logical 1s. Messages for a particular group of stations have addresses which begin with '1' and are then defined by whatever convention the system uses. All messages for individual stations use addresses which begin with a '0'. The rest of the 'frame' of data contains the address of the source of the message, its type, the message itself and a check sequence.

In a baseband system, any message on the network occupies the network completely. So each Ethernet node is also equipped with 'carrier sense', a detection circuit which warns if another device on the Ethernet is already transmitting. If the network is unoccupied, the node transmits. If something else is transmitting the would-be sender waits until the Ethernet is free. In a carrier sense, multiple access (CSMA) network there is no central controller which decides who transmits and in what order. In other words it provides distributed control of the network. The effect is like a discussion in a bar: no-one chairs the meeting so the participants have to grab the attention of the others if they want to say something.

In a CSMA network – also called a 'contention' network – it is possible for two stations to decide to use the same instant to start transmitting, causing a data collision. 'Collision detection' allows the colliding stations to carry on transmitting for a certain period – as long as it takes to transmit 32 to 48 extra binary digits – to make sure all the other stations know there has been a collision. After this 'jamming' period they stop transmitting and each waits a different random period – a different multiple of the time taken to transmit 512 bits – before trying again. The range of random wait times goes up with the number of collisions so that collisions can be resolved even if a large number of stations are colliding. This is what happens in the bar when, as will happen as the evening wears on, two or more participants try to talk at the same time. In such cases, the two will stop talking and try to begin again, repeating the process until one of the speakers gets in first. An Ethernet system allows 15 more attempts to retransmit. If transmission is still unsuccessful, an error is reported.

Ethernet has a lot going for it, which is why it has been chosen as the basis of the Technical and Office Protocols. It is a proven technology, now ten years old; the lack of a centralized network controller means the bandwidth of the network is used more efficiently and that, in most circumstances, the messages get to their destination without too much delay; the system is reliable because any one station's malfunction does not affect any of the others; and the system can be easily expanded just by adding new stations to the network. It has been widely adopted, albeit in a bewildering variety of slightly different forms. And though a full-blown Ethernet cable costs $20 a metre there are so-called 'thin-LAN systems' which run on coaxial TV cable costing little more than one-hundredth of this.

But, according to GM and its collaborators in the MAP programme, the possibility of delays caused by collisions makes Ethernet inherently flawed, particularly for industrial systems. Such delays mean the minimum time within which one node will be able to transmit to any other cannot be guaranteed. In other words, Ethernet is 'non-deterministic'. And in theory once the network is being used at a rate of 80 per cent or more of its capacity the collisions, and subsequent retransmissions, of data become so frequent that the system becomes unusable. It is out of the question for real-time control of, say, a boiler, robot, or other potentially dangerous plant because when alarm signals are being sent in some sort of emergency the whole system slows down – just when it needs to act, if anything, faster than normal.

This apparent problem is made worse because one of the characteristics of electrical connections is that, if a signal is travelling along a channel of any kind and suddenly meets a device – another channel or a computer or anything else – which has a different characteristic impedance or resistance to the travelling signal, the mismatch will cause a reflection which travels back down the line in the direction from which the original signal has just come. This reflected signal can cause collisions of its own, so impedances have to be carefully matched, which means expensive interfaces to the cable, and the need for high quality, expensive cable throughout. Ethernet is also limited in distance to 500 metres for any one cable section.

Before going on to describe GM's preferred method of getting round these apparent drawbacks to Ethernet it is worth pointing out that the alleged 'non-deterministic' nature of Ethernet is still a matter of fierce controversy. To begin with, there are many experienced users of Ethernet who claim that, in everyday use, the traffic on an Ethernet system rarely rises above five to ten per cent of its theoretical capacity. There are those, too, who have used Ethernet-type (CSMA/CD) systems in factories – the environment to which they are supposed to be unsuited – without any apparent ill effects. British Aerospace, as is explained elsewhere (Section 7.8), uses Ethernet for real-time, drip-feed programming of a number of machine tools, which is about the saltiest application a communications system can have, though it is worth noting that British Aerospace 'makes sure' that traffic on any one Ethernet segment does not rise above ten per cent of capacity. This suggests two things: that traffic might rise above this figure if not checked; and that ten per cent is regarded as the maximum permissible to avoid problems.

It must also be said that some of the distances claimed as a 'limit' for CSMA/CD have little relevance for the average company: few companies other than General Motors could wish a single length of backbone cable to travel 20 or 30 kilometres or more. Much of the argument has been conducted by vested interests, a fact which does not help the ordinary user to see the issues clearly. General Motors, for example, already has a considerable investment in broadband systems, so it could be argued that it is in GM's interest to promote the exclusive use of broadband, where possible, in order to create a volume market for broadband and so drive its own costs down. The main, or loudest, Ethernet protagonist, however, is Digital Equipment, whose president, Ken Olsen, has vociferously denounced criticism of CSMA/CD as ill-informed. Digital Equipment was an early developer of Ethernet and has a considerable stake in its continuing success.

These arguments are a pity, because to some extent they miss the point. As we hope becomes very clear later on in this book, the main functionality of MAP – TOP is outside the argument because it uses CSMA/CD anyway – has little to do with whether it uses CSMA/CD or not. There is considerable pressure, especially in European companies with different needs from General Motors, to incorporate MAP's features in a system running on CSMA/CD. However, that should not blind potential users to the great advantages of broadband. These are explained more fully in section 4.8.4, but the main one is that a broadband system, once laid, is likely to be able to accommodate all a company's future needs. That said, the controversy is unlikely to abate in the near future.

3.3.3. Token passing

The preferred method in MAP for getting round Ethernet's supposed problems is token passing. Token passing means that any station that wishes to transmit has to wait until a single token, or specific information packet, comes round to it. The equivalent is a board meeting, at which the chairman decides who speaks next and for how long. Thus only one node, that which holds the token 'key' to the network, may transmit at a time, and that for only a limited time. This means the maximum time each node has to wait before it can transmit is determined by the number of nodes on the network, not how heavy the data traffic is. Collision detection is not used so impedances and cable characteristics do not matter so much.

Token passing systems can work on a bus or a ring. In a token ring the token passes from one node to its neighbour and so on. The distances betwen successive nodes are short so the driving power needed to get the signal from one node to the next is minimal. In a token passing bus the token goes through a list of the nodes in order and, when it reaches the bottom of the list, it starts at the top again. The nodes are not physically arranged in a ring but the effect of this continually repeated sequence of nodes is to create an equivalent or 'logical ring'. The logical ring is determined by descending numerical order of active station addresses. If a system of priorities has to be established the logical ring can include some nodes more than once or some nodes may hold the token for longer than others. Receive-only stations can be on the LAN to listen but not transmit: they never get the token.

The access time to the bus is not determined by the traffic on the bus but by the number of nodes connected to it. MAP uses a token passing bus: the MAP token is a eight-bit control frame with the sequence 00010000. The token bus standard adopted for MAP is that drawn up by a committee of the US Institute of Electrical and Electronics Engineers (IEEE). This standard is IEEE 802.4 and the IEEE token ring standard is 802.5. More on the 802 committees in a later chapter.

Token passing systems either need a controlling station to monitor the stations for failure and to generate the token signal code or they must distribute these functions round the network, as MAP does. Wherever the controller is, it must make sure that important stations get the token as often, and for as long, as their seniority justifies. The token must be unique. If it is lost because of a transmission error corrective action must be taken. A master station can be instructed to regenerate a token if it does not see one within a specified time-out period but there is no need for token management to be undertaken totally centrally: in a distributed system each station or node may regenerate the token. In a MAP system the node waits until the token arrives, changes the pattern of zeros and ones in the token to a pattern known as a 'connector' which tells the other nodes a message follows, transmits its message, then regenerates the token and adds it to the end of the message. So the greatest challenge in token networks is to make sure that the token is not lost by an error, or that a node does not refuse to release it, sending out large amounts of data on to the network. Each station only knows the station that had the token before it and the station to whom it must pass it on. Therefore it is possible to have an error which calls up a station which is not on anyone's list, or passes it to a station that is not there. If a fault occurs, two stations might get the token at the same time. And what happens if the token disappears entirely? (The answer is that everyone shouts at once and an algorithm sorts out the problem.) All this means that error recovery is one of the most important aspects of the use of a token passing system. What is more, error recovery is distributed to each station on the network.

The error-recovery mechanism is as follows. When a node on the network has either finished sending its message or its allotted token holding time has expired, it sends the token on to the next station in the logical ring, that is to say, the station with the next lowest address. The node with the lowest address sends it to the highest address and the token works its way down the list. Consider two stations with addresses ten and nine. After station ten has sent the token to station nine, station ten listens for an indication that its successor station, number nine, has received the token. It does this by listening for a valid frame. A valid frame must mean that station nine has started transmitting, so station ten assumes that nine has received the token. If it does not hear a valid frame, it listens for noise and passes the token – 00010000 – again. If the second attempt elicits no response after a timed period, station ten assumes nine has failed and sends the rest of the network a 'Who follows?' frame, followed by station nine's address. Each station in the network has a source and a destination address plus a means of determining where the token goes next. Thus the station that would have succeeded nine compares the address after the 'Who follows?' frame, recognizes it and sends its own address in a 'set-successor' message to ten. Station ten then sends the token to the new successor station, deleting the failed nine from the ring.

If the 'Who follows?' frame elicits no response, ten sends it again. If this produces no response ten sends another 'solicit successor 2' frame containing ten's own address in the destination address and source address fields. Any station which receives this frame and needs to be part of the ring will respond, and the ring will be re-established. If nothing happens, the station with the token assumes that:

All the stations in the ring have failed
The sending station's receiver has failed
The medium has been severed

and it then stops trying to keep the ring going.

One other type of token passing system should be mentioned, the 'slotted' ring, of which the so-called 'Cambridge ring' is an example. The 'slot' is the data equivalent of an envelope. And one or more envelopes may circulate in any one ring system. A node which wishes to transmit waits for the next available empty envelope to pass. It puts its data into the envelope and adds the address of the destination in the appropriate part of the frame. The envelope goes round the ring and, when it reaches its destination, the node copies the data and adds a note on the envelope to say the data has been received. The envelope continues its journey and eventually reaches the node which sent the original message. When the sending note sees the note from the destination node it 'empties' the envelope and sends it on its way.

3.3.4 Broadband and baseband

As already mentioned, MAP uses token passing and Ethernet is a contention network. MAP differs from Ethernet in another important respect: Ethernet is baseband, MAP is broadband. This is because, when General Motors formulated MAP, they considered two things important: one was that the network should be deterministic, hence the use of a token passing bus; the other was that the network should make use of the broadband cable systems GM had already installed.

GM was already used to broadband, which can carry speech and TV as well as factory floor data. And, thanks to the 40 million US homes connected to broadband community access or community antenna TV (CATV) stations, broadband equipment was, and is, cheap and proven. GM decided, after taking advice, to make MAP a broadband, token passing bus based on the International Standards Organization's open systems interconnect (OSI) model.

Digital networks break down all information into logical 0s and 1s. In baseband systems the 0s and 1s are represented by high or low voltages which take up the whole transmission medium, cable or whatever, for as long as the transmission takes place. In a broadband system, the signals can be transmitted by large signal and small signal bursts of a higher frequency signal. If a number of different high frequencies are used, more than one stream of data can be sent along the medium at the same time. A variant of the large-signal and small-signal – amplitude modulated (AM) – coding is used in full MAP broadband.

An alternative is to use two frequencies, one to denote a logical zero and the other for the logical one: in this 'frequency shift keying' (FSK), the two frequencies are based on a single frequency which is shifted up or down to create the zero and one signals. This is the principle used in

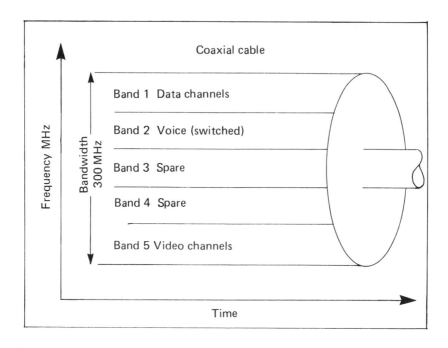

Figure 3.6 Frequency division multiplexing of a single transmission medium into several broadband channels.

the so-called 'carrier band' versions of MAP, about which more will be said later (section 4.8.2).

In a broadband system as many different sets of signals can be sent along their own frequency specific channels as can be accommodated by the full bandwidth of the cable. The obvious comparison with this frequency-division multiplexing (FDM) (see Figure 3.6) is the radio set, which receives all the radio signals available to its aerial but which is tuned to select the particular channel the listener wants to hear.

Broadband coax offers much greater bandwidths than baseband coax, some 300 to 400MHz. The bandwidth of a transmission medium can be broken down into multiples of the frequency band occupied by each channel. So this means a single 350MHz cable can accommodate one 350MHz channel, or two 170MHz channels (with a 10MHz 'guard band' to keep them apart and prevent interference between the two), or ten 34MHz channels with a 1MHz guard band between each channel. This means broadband can accommodate simultaneous transmission of data, voice and vision. On the kind of broadband systems we are

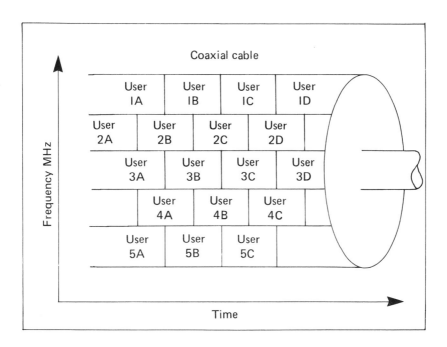

Figure 3.7 By combining FDM and TDM, many more users can share a single transmission medium

concerned with here, individual channel data rates are normally from one to five Mbit/s, or two channels can be combined to give 10Mbit/s, and it is at that sort of rate that the broadband systems become economic.

A broadband cable can carry even more channels than its bandwidth allows if, in addition to FDM, time-division multiplexing (TDM) (Figure 3.7) is used. TDM allows several users access to a channel in quick succession. The signals are effectively broken up and labelled so that the different components of a message can be reassembled by each user later on. Broadband has the advantage that it can apply TDM to one or more channels and offer Ethernet, token passing, RS232, video or voice on the others. This means it has great potential for teleconferencing, training and other uses.

Indeed, one of its main attractions is the potential uses it offers. As mentioned later, broadband is more expensive than baseband – about five times more expensive – but its installation can often be justified because it will be able to cope with all the extra future demands any company is likely to make of it, for all sorts of uses. Broadband also offers high immunity to electrical noise and can span the long distances typical of factory and company-wide installations.

That said, although broadband cable itself is not that expensive, using even CATV type broadband devices in a factory can be, in the early stages. Each device that needs to send messages along the network needs a modulator to get its message on to one of the broadband channels at the available frequency. It will also need to receive signals, so it needs a demodulator to carry out the process in reverse. Since most devices need both to transmit and receive, each needs a modulator/demodulator ('modem' for short) to connect to the network. They are not cheap.

Using token passing on a broadband system means priorities in the network can be more easily controlled. But this also means that, in a distributed token management system like MAP, the cost of network control circuits must be added to the modem-per-station cost.

So the need for data modems at each node on the network to send and receive information plus high capacity coaxial cable makes broadband expensive unless lots of information is being sent on the network.

3.4 Head Ends

The signals discussed so far are at radio frequencies (RF). RF signals are directional and the easiest and cheapest way of using RF signals on a cable is to arrange that nodes pass their signals on to the broadband bus in one direction and receive them from another. This is the way the RF signals are used in CATV equipment. No node or station will receive a message unless the message is sent along the cable at the correct 'receive' frequency. Every node must transmit at the correct 'send' frequency.

When a node receives its token, it transmits its data by impressing the signals on the 'send' channel. The technique used to do this in MAP broadband is a combination of amplitude-modulation (described above) and phase shift keying (AM/PSK). This uses a combination of signal-level and phase information to increase the data rate – the number of bits per baud (see Section 3.2.1). The signal carrying the data travels to a head-end remodulator ('head end' – Figure 3.8) which transfers the data to the 'receive' channel and puts it back on the cable in that form. All the nodes can then listen for it.

This means that all the signals must go through the head end before they can be received. This is one weakness of a MAP network. It means that, if the head end fails, the whole network fails. This proved to be a constant problem in preparing for the CIMAP demonstration in

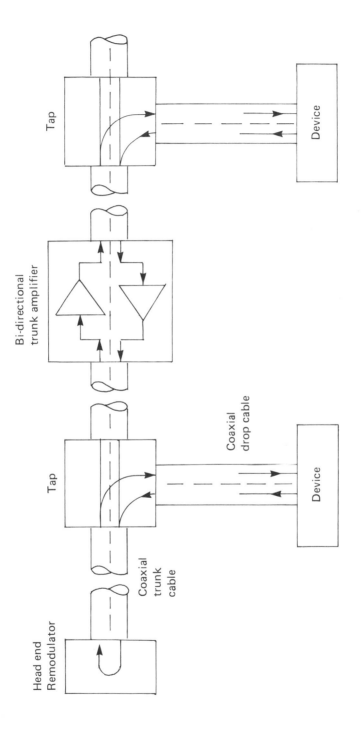

Figure 3.8 Frequency multiplexed cable system with amplifiers and head end

England, for example. It can be solved by having a 'redundant' head end on 'hot stand-by', which means that a spare head end is always powered up ready to take over.

Note that, apart from the need to make sure the head end is functioning correctly, the actual details of the way the head end operates need not concern the user. As far as the user is concerned, the token passes from node to node in a logical ring without having to return to a headquarters between each call.

The network characteristics outlined so far – topology, baseband and broadband, modulation, methods of access to the network and so on are basic considerations in communications. But the way the network is used to do a real job of work is a more complex matter. In fact, sorting out such matters takes up only the two lowest layers of the seven layers in the OSI model which was developed to identify and deal with the complexities of communications. The seven layer model is described in the next chapter.

4
THE SEVEN LAYER MODEL

More needless mystery probably surrounds the Open Systems Interconnect (OSI) seven layer model than any other aspect of MAP and TOP networks. The definition of the model is itself an international standard (ISO 7498), drafted by a committee – technical committee TC97, sub-committee 16, or TC97/SC16 – set up by the International Standards Organization (ISO) in 1977.

The model is necessary because otherwise it would be extremely difficult to arrive at a general description of the communications process which covered the behaviour of every type of communications system. The way a phone network operates is different from the way a factory local area network or a satellite link behaves. Each has a different way of managing connection, transmission and disconnection in a communications session. A communications system has to carry out so many tasks to manage a session – and many are going on at once – that these tasks are only manageable if they are subdivided. Previous so-called communications standards have provided only incomplete definitions of the way information can be transferred, leaving the gaps to be filled in non-standard ways. The OSI model is a general description of the communications process: it attempts to identify and define all the tasks which might need carrying out in the communications process and subdivides these tasks among the seven layers of the OSI model. A main part of the value of the OSI model lies in the completeness of its definition of the communications process.

The tasks at each layer are defined and carried out by protocols, or rules, specific to each layer. The OSI model only describes the functions which each layer must carry out: it does not establish which protocols each layer must use. The choice of protocols is left to each particular OSI implementation. MAP is an OSI implementation which uses a specific set of protocols and TOP is another which uses a set of protocols slightly different from those selected for MAP. One big attraction of using the OSI model is that gateways can be used to convert the protocols selected at each layer for one OSI implementation into those used at the equivalent layer of a different OSI implementation.

Not any protocol will do. The protocols available for each layer are defined by particular OSI international standards specified by the ISO. Because each layer is independent of those above or below, though communicating with them, it is possible to begin with a standard for a particular layer function, then, when the standard is overtaken by technological change, to update it without affecting the operation either of that layer or of the rest of the model. This is one of the most important features of the ISO model, and explains why it is dominating most of the talk about standardization.

A complete set of ISO standard protocols is not yet available for all the layers, though sufficient progress has been made to say that most of those needed will soon be available. The standards selected at each layer for MAP and TOP are shown in the diagram (Figure 4.1).

The bottom two layers are implemented in hardware but all the third and higher level functions can be realized in software, depending on which protocols are chosen. This software can be either built into integrated circuits or stored by integrated circuits which accept the software as a download from a host system. This download is accomplished when the system is started up. In either case, software changes can be made as the protocols for the higher levels become firm standards. Chips supporting the standards can be made in volume very cheaply, one of the main benefits of standardization.

The model identifies what has to happen if information has to be exchanged between two systems which are linked over an unspecified communications network of some kind. The division of the functions so identified into seven 'layers' is arbitrary and in fact, the further you examine the way the model is used in a practical system, the more obvious it becomes that some of the so-called seven layers divide into sublayers and others may not be needed at all.

Some functions, such as addressing, are a multi-layer problem. Addressing is necessary for locally identified names processed by the session layer, global addresses processed by the transport and network layers and the physical layer addresses processed by the datalink and physical layers. Network management is not specified in the OSI model at all and both MAP and TOP have to define their own until the ISO formulates this.

The seven layers can be grouped, as well as divided, in various ways. The top three layers of the stack, for example, serve the application,

Layers	TOP 1.0 Protocols	MAP 2.1 Protocols	Function
Layer 7 Application	ISO FTAM (DP) 8571 File transfer protocol	ISO FTAM (DP) 8571 File transfer protocol, manufacturing messaging format standard (MMFS) and common application service elements (CASE)	Provides all services directly comprehensible to application programs
Layer 6 Presentation	NULL (ASCII and binary encoding)		Transforms data to and from negotiated standardized formats
Layer 5 Session	ISO Session (IS) 8327 Basic combined subset and session kernel, Full duplex		Synchronizes and manages dialogues
Layer 4 Transport	ISO Transport (IS) 8073 Class 4		Provides transparent reliable data transfer from end node to end node
Layer 3 Network	ISO Internet (DIS) 8473 Connectionless and for X.25 – subnetwork dependent convergence protocol (SNDCP)		Performs message routing for data transfer between nodes
Layer 2 Data link	ISO logical link control (DIS) 8802/2 (IEEE 802.2) Type 1, Class 1		Detects errors for messages moved between nodes
Layer 1* Physical	ISO CSMA/CD (DIS) 8802/3 (IEEE 802.3) CSMA/CD media access control, 10Base5	ISO token passing bus (DIS) 8802/4 (IEEE 802.4) Token passing bus media access control	Electrically encodes and physically transfers messages between nodes

*Note that ISO is considering moving the IEEE defined media access control (MAC) sublayer of the data link layer to layer 1, the physical layer. This move would make the MAC sublayer concept conformant with the OSI reference model.

Figure 4.1 MAP and TOP common core protocols

while the aim of those below is to ensure reliable communications, but other groupings have been offered: This is why it is best to treat the model as a series of questions each system asks the other about the rules the two systems will follow before a data exchange. In effect the two systems negotiate the mutual rules before application data is exchanged, and labels attached to the data ensure that the rules are followed. An address label, in fact would provide a good analogy were it not for the fact that the word 'address' has obvious data communications connotations and the fact that there is much, much more to the seven layer model than a series of labels. But the analogy does work in the sense that a complete address may be written in the form:

>Ms J Smith
>Flat 4
>Tower House
>Littlehampton Street
>Middletown
>Barsetshire
>England

An envelope bearing such information might be posted anywhere in the world. If the letter is posted air mail in the mid-Western USA, the first person to see it where it is is posted, in Kansas or wherever, is not interested in the name, the flat number or anything else but the bottom line of the address. When the letter reaches England the first person to see it there is not interested in the 'England' part of the address at the bottom, nor in the top five lines, only the part which narrows down the destination in the next logical way, to Barsetshire. In Barsetshire the person sorting letters is interested in the 'Middletown' line and so on. When the letter goes through the door of Flat four, its occupants want to know which of them owns the letter. It is only when this is established by the very top line that the letter reaches its destination. Note that none of this has any bearing whatever on the contents of the letter. It is up to the sender to make sure the recipient has something to read and that the contents can be understood.

In the same way, the seven component layers in the seven layer model are interested only in the layer above and the layer below. The active elements in each layer, known as entities, are those hardware or software elements in the layer which carry out particular functions in that layer. These functions are carried out by adding to, or reading from, information on the 'envelope' in which the message is carried. It is as though the address shown above were added line by line as the letter progressed on its journey until it reached the aeroplane which takes it airmail to the country of destination. When it reaches that country the address is peeled off line by line until only the name of the recipient is left.

The seven layer model is based on the idea that a seven layer stack exists at the sending end and at the receiving end. The entity in a particular layer, call it layer 'n', at the receiving end, has a particular interest in the information provided by, and the service or action required by, the layer 'n' entity at the sending end. The layer 'n' entity, in other words, acts upon the instructions or advice of its peer entity at the other end of the communications channel. These instructions are contained in binary encoded data which the peer entity has added to the data as it has come down the stack at the sending end. The data is contained in a protocol data unit (PDU). PDUs are usually referred to by the name of the peer layers between which they are communicating, for example, a transport PDU (TPDU) or session PDU (SPDU). Each layer adds its own data and/or control information to form its own layer-specific PDU, so that only the bottom-most layer – the communications channel itself – sees the equivalent of the whole 'address' shown above.

The way all this works has been well compared by writer Michael Witt with the peer-to-peer communications between presidents in two corporations who want to arrange a softball game. The president of one corporation asks his secretary to arrange a series of games. The secretary talks to the softball team captain, who then phones the softball captain in the rival corporation. The preliminary communication might be along the lines: 'Are you interested in a game?' If the answer is yes, the conversation will turn to when and where the first of a series of games might be played. All this would have to go back up each hierarchy until both presidents of the corporations were satisfied with the arrangements. The only communication channel used would be the phone line between the two captains. But the effect would be that the two presidents were communicating, peer-to-peer.

At each layer, the seven layer model adds protocols to the basic applications data. These protocols deal with the needs of a particular application, without knowing its contents. The equivalent, in terms of mailing a letter, would be such matters as 'Is this air mail?' 'Yes? Then put an airmail sticker on here.' It asks similar questions about whether the letter is formal or a personal letter, what colour and size the envelope and the paper the letter is written should be. How far has it to be sent? How much does it cost? Is it pre-paid or is a stamp to be used or will it be paid for on delivery? Should it be registered or not? Must the recipient acknowledge its receipt? Should the letter be scented or unscented, typewritten or handwritten and, at the most basic level, what typeface should be used?. Note that none of this affects the words used: when all the questions about the issues surrounding the sending of the data have been answered, the writer is still at liberty to begin a job application 'My Darling Harriet' – with predictable results.

From theory to practice. The seven layer model as outlined above is made up of different OSI-approved specifications and, since each layer is only concerned with

1. the specification defining the services it offers to the layer above, and
2. the protocols used by or the transactions between the two systems which will allow it to provide those services

then each layer can mix and match different specifications for different purposes to achieve the desired spread of services. Currently there are over 80 possible standards, draft standards, draft proposals or working drafts available covering all seven layers.

Layer entities use PDUs to communicate with entities in equivalent layers in other parts of the network – like the secretaries in the corporations above. Each entity exchanges service primitives with adjacent layers in order to

1. provide a service to the layer above and
2. use the service provided by the layer below.

The exchange of information – PDUs – between corresponding layers in two connected systems is possible only because each layer in each system provides services to the layer above. The information passed and the services which enable it to be passed are separate, though connected.

The information passed is in two parts:

control
data.

The *control* information is the basis for all the services that are required to process the message. As each layer provides its part of those services, the remaining control information is passed to the next lower layer.

With one exception, the *data* is usually passed down to the layer below unchanged. The exception is at the presentation layer, layer six, which may reformat the data. As the data goes down through each layer it is prefaced with control information in a control 'header' before requesting services from the next lowest layer. This control information is interpreted by the corresponding layer in the seven layer stack in the system which receives the message. The addition of control information means that the size of the block of data which incorporates the message grows larger as it goes down the seven layer stack, reaching maximum size as a complete frame when it is passed through the physical layer on to the network. At the datalink layer control information is added at the end as well as the beginning of the frame. The control information is stripped off, piece by piece, in the receiving system. As already touched on, at the various points in the complete frame's construction, the frame is called a protocol data unit (PDU), which includes control information, address information and data. Like the seven layer model on which MAP and TOP are based, neither MAP nor TOP is concerned with the internal operation of the systems they interconnect, only the fact that they interconnect.

4.1 The application itself

The model begins with the application, which is above or separate from the model. Every application will be transferred in 'frames' of data which must be organized so that the system at the other end can receive and understand them. *The job of the communications system is to transfer that frame, not to reinterpret it for the receiving system.* The frame may be broken up and reconstituted to suit the communications medium on which the recipient system is sitting. For example, the frame may be sent from a system on MAP to a system sitting on TOP or some other Ethernet-type system. But the frame must, after the communications process, be understood internally by the recipient.

4.2 Applications layer (layer 7)

The top layer of the model is the applications layer. Its job is to provide services for the application. The available services vary widely but they could include identifying the source of the data and its destination; whether the two are ready to communicate; establishing whether the data is authentic and if the communication is authorized; establishing any privacy mechanisms which may be needed; establishing how the communication is to be paid for, whether the money is there and what level of service can be provided for the stated amount; making sure that the exchange of data is consistent; agreeing the rules for starting, continuing and ending the communication; agreeing some data syntax, such as the available character sets; agreeing responsibility for error recovery.

All these are general functions of the sort that an application might expect to support it. The application layer in general has two parts. There are common services which most applications will need access to. These are served by what used to be called the Common Applications Service Elements (CASE). CASE manages the service elements in a particular communication and makes sure that high integrity interworking is available when needed.

4.2.1 ACSE, formerly called CASE

The entity in an application layer consists of a user element, a CASE and a SASE. In 1986 the CASE sub-layer was renamed ACSE, which stands for Association Control Service Element, because this name more accurately reflects its function, ACSE provides facilities which control the association between the application entity in layer seven of one system with the application entity that it wants to communicate with in another system. It provides services to establish and terminate application associations. The SASEs provide services to the user element.

At the application level each application entity must know the title of the far application entity, the application SASEs the other entity uses, and the abstract syntax the other entity uses. The application entities are identified by the presentation service addresses they use. The applications entities obtain this information during the early part of a communication by sending particular request and response 'service primitives'.

In the minimal ACSE kernel specified in MAP there are three types of service primitives: one to set in train the exchange of association information, another to release the entities from the association at the association's end without the loss of any data and a third to abort an association before it has finished. ISO specifies different versions of these primitives for different functions: request, indication, response, confirmation. In the case of abort, however, only request and indication variants are used.

The SASEs are specific to certain applications. Examples include specific support for industries such as banking (credit checks and the like), airlines (reservations, timetable enquiries and so on).

4.2.2 FTAM

In manufacturing, both MAP and TOP use the File Transfer and Access Management (FTAM) protocol. This is ISO draft proposal 8571. FTAM gives remote access to files held on different minicomputers, whether the minicomputers are on the broadband factory back bone or on an 802.3 network elsewhere in the operation. Exchanging messages between computers and controllers is the job of the manufacturing message service.

Consider two minicomputers, A and B. If A is the FTAM initiator and B the FTAM responder, a user application at A requests its FTAM application entity to operate on a file that resides or is to be created in B's file store. The MAP or TOP network then transmits this request to the FTAM entity at B. B's FTAM service user then accepts and acts on the request and does the appropriate file operation. The user need not know anything of how this is done. In fact it is FTAM operating as a virtual file store.

FTAM guarantees to create, delete or transfer files, read or change file attributes, erase file contents, locate specific records and read and write records of a file. FTAM may also support sequential files, random access files, or single-key-indexed files. All that is required is that the files be classifiable under one of nine types defined in a National Bureau of Standards document, NBSIR 86-3385-3: unstructured binary files; unstructured ASCII; unstructured files which conform to ISO 8859; sequential ASCII; same but with 8859 text instead; sequential file types; random access file types; indexed sequential file type with a single key; and file directory type. These are numbered NBS-1 to NBS-9.

FTAM defines nine functional units, each associated with a set of file services, such as read, write, recovery and restart data transfer. FTAM also has five overlapping service classes: file transfer, file management, file access, file transfer and management, and unconstrained.

Although FTAM allows the user to transfer complete files, to access and transfer parts of files – for example, records – and to read and manipulate file attributes, many FTAM implementations are confined to file transfer, not file access. File transfer allows the receiving terminal to see the data but not to change it, whereas file access means altering a file and putting the altered version back in place of the original.

Note that FTAM is still under development. The interfaces to the FTAM services are unspecified, for example. This means that the definition of an interface to the FTAM services is left to each individual implementer, and though the drafters of the preliminary MAP and TOP version 3.0 specifications had hoped to include an FTAM application interface, the text of this was still under public review when the drafts were published in the spring of 1987. The texts were expected to be published as this book went to press.

4.2.3 MMFS and MMS

The MAP application layer contains both a file transfer service, FTAM, and a messaging service. FTAM's main use will be to transfer files between minicomputers connected on the factory backbone – the main broadband bus. But the messaging service also has a file-transfer facility to transfer files between controllers, minicomputers and other shop floor devices.

There are two types of messaging service – the manufacturing message format standard, MMFS, which is used in MAP 2.1 systems – and its later variant, the manufacturing message standard, MMS. The development of these variants and the differences between them are outlined in Chapter 6. For the moment it is important to concentrate on what the messaging service does.

MMFS and MMS provided a standard syntax for messages between factory devices. They assemble frames of data into complete messages which can be passed between numerical controllers, robot controllers, programmable controllers, automatic guided vehicles and computers of various types. An MMFS message, for example, consists of a series of fields, or complete data blocks which, taken together, form one of a number of MMFS protocol data units. MMFS defines what the fields should contain and in what order the fields should be assembled in each PDU. Certain parts of the MMFS syntax apply to certain devices – programmable controllers, robots and so on – and each particular device, whatever its type, is loaded with the minimum sub-set appropriate to its type.

MMS also assembles frames of data into complete messages and defines the message notation. This is important because the message notation contains information about the purpose, length and value of each element in a message. MMS describes the range of messages available both to the application processes and to other SASEs. And it contains information about what should happen when a message is received.

It is difficult to stress too highly how important these messaging services are to the MAP specification – they are not used in TOP. Without MMS (we will concentrate on MMS since this is the chosen messaging system in MAP 3.0), MAP is just another communications system. MMS makes it much more than that.

MMS is the key to achieving vendor-independent interoperability between shop floor devices. MMS exchanges pairs of messages – one in each direction – between any computer on a MAP network with any factory item connected to the network. It allows cell computers to exchange files either with one another or with a machine tool controller, a robot controller or programmable controller running a conveyor. The cell computers could use such a message to alter a robot's control program. Or the computer can ask the controller to use an MMS sub-routine to ask another device for information which, when the second dialogue is over, can be read back up to the computer which made the first request. The service allows a number of important factory floor functions such as remote control, access to user or part-program files and so on.

There are three ways MMS loads user programs into programmable devices on the shop floor. As explained further in Chapter 6, two of these control the devices from a host computer which must know about the devices it controls. This method is not vendor independent. In the third method, the host needs no information about the programmable device on the shop floor and does not even need access to the user program. It uses MMS to send a message to the device – which must be MAP-compatible – telling it to request a program and leaves the device to find the program in a hard-disc store, a remote database or its own memory, and run that program.

The load can be either the whole program or part of it at a time, so-called 'drip feed'. So, using MMS, even simple machine tool controllers could supervise long, complex machining routines involving multiple cutters, probing cycles and so on, regardless of the limits imposed by the machine-tool controller's own memory.

As for other services, each MMS entity communicates with its user element by exchanging 'service primitives' with that element. And, similarly to other such entities, an MMS entity communicates with the MMS entity in the other device on the network by exchanging MMS PDUs. There are four main types of MMS PDU – request, response, error, and reject – applicable to MMS codings and four types of service primitive – request, indication, response, confirm.

If the user element requests an MMS service, the MMS entity provides a protocol data unit for transmission to its peer MMS entity in the other stack. For example, confirm or response primitives are positive if a request for an MMS service is successful and negative if it is unsuccessful. This means that, if an MMS entity receives a response positive primitive from its user entity it will transmit a response PDU. But if it receives a response negative primitive it will transmit an error PDU. Reject PDUs are transmitted by the MMS entity if it receives a

PDU it does not recognize or believes some other error has occurred.

There is a large number of available MMS codings providing all the functionality mentioned above. So far, the best examples we have of these are the codings used for the CNMA project (see section 7.8). They include initiate, conclude, abort, cancel, identify, read, write, file open, file close, file read, obtain file, load from file, store to file, start, stop, status and so on. Each of these would be qualified by the generic PDU types above, so that the messages would include:

File open request
File open response
File open error

and so on. Each of these is allocated a maximum length.

Devices have available some services and not others. The control computer might read files in another computer using file open, read and close. The obtain file service prompts another computer to read files. The load from file service gets a program ready for execution in a controller. The store to file service stores a controller program in the control computer and the status, start and stop services are used to control machine tool or programmable controllers. An NC, robot or programmable controller may use file open, read and close to obtain tool data or operating instructions. It can use an unsolicited status service to report to the control computer – although the control computer could achieve the same result by polling round each controller at periodic intervals, just as programmable controllers themselves operate – and a controller can read or write data variables in a computer or controller using the read and write services. A control computer can ask a controller to ask a third device for information which can be read back to the control computer.

In CNMA these messaging services were used to pass:

Order information to a central computer so that the work schedule could be updated;
Status information to remote computers for displaying cell status

As already noted, the control computer gives commands to the controllers. Four simple examples from CNMA are:

Load a pallet of new tools into the machining centre
Load a new billet for machining
Machine the component
Dispatch the new component to stores.

4.2.4 Other SASE elements

Other SASE elements include job transfer and manipulation (JTM) which allows work to be submitted on one open system and run on another. JTM could allow files to be collected from a number of open systems outside the immediate one and the results of the manipulations on those files to be delivered to other systems, perhaps for further processing. JTM provides a way of implementing distributed batch jobs. It specifies the way users should specify the division of a job and where and when each part of the job is to be processed and delivered without the user needing to supervise these tasks.

Virtual terminal (VT), as defined in TOP 3.0, is a remote terminal access building block. VT manages the transfer of data between users on a variety of hardware running a variety of host applications. The service is supplied by VT application entities in the application layer. The two relevant standards are ISO 9040 and 9041.

The VT is represented by an object-based computer model. The structure of this object is a one, two or three-dimensional array in which each element may contain a single character representing any of four terminal characteristics: colour (screen background and foreground), emphasis, character repertoire and font. Interacting VT users can update this object to simulate the capabilities of a particular terminal. When an association is set up between by two VT users, the users initiate that association by specifying all the values they need for these elements. Users may specify various types of profiles, which are predetermined sets of values, by default, by using registered alternative profiles, or by specifying privately held profiles.

Each VT user has a set of objects representing the VT. The TOP VT protocol makes sure the sets of different objects agree with one another and carries out syntactical translations where necessary. TOP version three provides that separate display objects are needed for the display and the keyboard for each VT.

4.2.4.1. Directory services

Another set of services provided by the application layer is the directory. Changes are taking place in networks all the time. Nodes come and go, routes become available or unavailable, and nodes or networks can be relocated. A user may not need or want to know what changes are taking place in a network but the relevant nodes must keep themselves updated on these changes. The directory service makes this possible. There are two levels of directory service, one local and one global. Each node in a network will have to keep track of the changes at other nodes on the same network, but only one node in a network needs to keep track of change in the other networks to which a single LAN is connected.

The way a node keeps up to date is to keep a client service agent (CSA) which can access a look-up table of addresses. The table is updated regularly by a directory service agent. A central table is kept up to date by a network administrator.

4.2.4.2 Message handling system

The message handling system or service (MHS) is an important feature of the electronic mail building block specified for TOP. MHS allows the 'store and forward' exchange of text messages, which means that the information is stored at various points along its route until enough network resources are available to send it along the next leg of its journey.

MHS has been drawn up by the users and specifies two services: a basic message transfer service (MTS) which supports general application-independent services, and interpersonal messaging (IPM), which uses the MTS to support electronic mail. IPM and electronic mail are synonymous. Other services will be added to these, such as electronic data interchange, computer to computer dataset transfers and other non-electronic-mail applications.

TOP's MHS is based partly on the CCITT's X.400 recommendation, though it departs from this in certain respects: X.400 requires Class 0 transport protocol but MHS requires Class 4 (full error recovery, see Section 4.5.2). MHS is also based on the US National Bureau of Standards OSI implementation agreements.

The MHS is made up of a collection of message transfer agents (MTAs) and user agents (UAs). A UA is an application process that interacts with the MTS to submit messages. In other words, this is the word-processing system, personal computer or other terminal which creates the file of text which is to be sent. A user prepares messages using a UA and the MTS, in turn, delivers the message to one or more recipient UAs. The message is made up of an envelope, which carries the information, and the contents. The UA transfers the contents of the message to an MTA with a submission envelope which provides the information the MTS needs to provide the requested service elements. At the other end, the MTA transfers the contents plus a delivery envelope which contains information about the delivery of the message. At each stage, responsibility for the message is transferred between UA and MTA, and the message is relayed, store and forward, to a succession of MTAs until it reaches the recipient UA. As it is being passed from MTA to MTA the contents are carried in a relay envelope which carries information about routing, charges and so on. UAs are grouped into classes which define the type of messages they can handle.

The TOP MHS defines the envelope contents, which must be simple lines of ASCII-type text. Later versions will provide for binary, fax and other contents. These others include types which are being studied by TOP working groups and include office document architecture (ODA) (see Section 4.10), computer graphics metafile (CGM) and the initial graphics exchange specification (IGES), which specifies a neutral format enabling engineering-drawing data to be exchanged between different vendors' CADCAM systems.

The TOP MHS also defines a gateway which allows information to be exchanged between different proprietary mail systems (PRMDs). These are privately owned and operated MHSs which are grouped together for administration purposes. The TOP MHS also defines methods of interconnecting systems within any particular PRMD.

4.2.4.3 Network management

There is one more service which in theory resides in the applications layer but in reality applies to all of them. It is network management and it is responsible for operation planning. This includes gathering such statistical data as the amount of traffic and the number of errors occurring.

In a distributed network, network management is necessary to make sure that all the layers are interacting properly and are aware of what is going on in the network. If a system on a network develops a fault it should be arranged that other systems on the network first know about the fault and then can work their way round it using a routing algorithm.

Network management also allows the automated 'tuning' of some of the parameters of a network, say by altering timings at the transport layer. It might also carry out accounting management, which is not very well developed in the specifications at the moment.

To carry out network management so far MAP has used the fairly crude 802.1 'high level interface' recommendations, which are being developed into a full ISO standard. But network management is a problem area because ISO's progress in providing satisfactory standard solutions to replace the myriad proprietary solutions on offer has been slow. These functions can be carried out manually for short periods of time and this, indeed, is how some of the MAP and TOP demonstrations have succeeded in overcoming the shortcomings of MAP's 2.x specifications. At demonstrations there are always lots of highly-skilled people around who can make up for the lacking functions. But an automated substitute is absolutely essential for users who wish to make use of MAP or TOP networks without access to such resources. This is particularly true when the 'real' networks are much larger and need more management than the 'demo' networks.

4.3 Presentation layer (layer 6)

There may be three different ways of constructing the data in a communications system: the way the sending system constructs it; the way the receiving system constructs its own data; and the way the communications system between them constructs it for transmission. All three may be the same; they may not. The job of the presentation layer is to make sure the necessary translations are done. MAP 2.1 and TOP have skipped this process by making sure that all three languages are the same. For example, one of the differences between MMFS and its later variant MMS, for example, is that MMFS uses a non-standard syntax (language) instead of the ISO's Abstract Syntax Notation One (ASN.1). MMS uses ASN.1, which is a slight variant of the X.409 syntax recommended by the International Telegraph and Telephone Consultative Committee. It is at layer six that, for the exchange of text, there is either translation to or agreement about the use of the American Standard Code for Information Interchange (ASCII) for the characters that certain combinations of logical 1s and 0s represent.

4.4 Session layer (layer 5)

The session layer uses services provided by the layer below to establish how a communication between two users may be initiated, managed and co-ordinated, then ended. A 'user' for session layer purposes may be a terminal user or a program which is running in some system attached to the network. The 'kernel' service sets up and closes down the connection. This layer also determines which of three modes of interaction the communication will take. One is simplex, which means that only one of the two connected users may send data throughout the session. The second is half-duplex, which means that only one user can speak at a time, rather like a radio-telephone. This is done by the exchange of tokens – not the same as the broadband tokens used on the MAP network – and the session layer manages the exchange of these tokens. The third is full duplex, which means either can transmit at any time and be received by the other.

The session layer can arrange 'expedited data exchange', which means a selected amount of data is sent to the front of the queue for transmission ahead of anything else: it may even 'overtake' data already transmitted. A 'typed data' service allows control information to flow against the data direction established by the token in half-duplex mode. Another important service the layer offers is synchronization: users can place marks in the data flow to mark and acknowledge identifiable points in the communication. Should an error occur, both users can agree to go back to the unique serial number of a particular mark. The serial number is managed by the layer but its assignment is left to the users. This is important should the session layer lose the use of the transport layer below it at any time. A series of transport connections may support a single session, or one transport connection can support a series of sessions.

4.5 Transport layer (layer 4)

The transport layer carries huge responsibility for the success or otherwise of the transmissions in a data network. It is responsible in particular for error correction, about which more shortly. But in general it is the layer which establishes connection between applications running on two distant systems independently of the nature of their physical connection or the subnetwork they are on. The transport layer provides an 'end to end' communications path, right from one communicating system to another. It is this layer which provides the session layer with the means to establish, maintain and break off a connection.

One way to split the seven layers in the OSI model is into the bottom four layers, which are concerned solely with the communications aspects of the data exchange, and layers five to seven, which are concerned with the applications which will be available to the user. So if the top three applications-oriented layers are riding on top of layer four – the transport layer – it is clear that layer four must provide the important overall communications functions which ensure that the application will work, unimpeded by any failings in the communications system.

4.5.1 Connection-oriented and connectionless communications
This means in effect that the transport layer must provide the 'connection-oriented' transport service. Networks may be either connection-oriented or connectionless. A connection-oriented network uses three phases of connection: it establishes the connection; exchanges the data that needs to be exchanged; then gracefully arranges the disconnection. No data exchange takes place unless a correct

connection is established first, just as happens when a phone call is made: unless a phone caller hears the other party pick up the phone, no data – speech in this case – is exchanged. And in a connection-oriented network there are constant checks to make sure the call is still getting through right from one end to the other. It is the job of the transport layer to provide this connection-oriented service.

Connectionless networks use only the data transfer phase, that in the USA are called datagrams: communication begins without the certainty of a connection. The packet of data sent out carries with it all the information it needs about priority, addressing and so on to route itself through to the recipient. If the packet does not get there this has to be sorted out some time later, just as in the postal service someone might wait a couple of days before complaining that a letter which should have arrived did not.

Connectionless systems may be quicker but, if they are to work, they must be used on networks which are highly reliable. Reliable or not, local area networks are inherently connectionless: nodes transmitting on a LAN send out data without checking that the recipient node is ready, though there is some interaction for flow control (see below).

There is much more to the connection-oriented service the transport layer provides than the preliminary exchange of a simple acknowledgement, what is called 'handshaking' in telecommunications. There is an exchange of state information, that is control information, which is basically is being exchanged at the same time as data transfer. There are also various negotiating functions. One task of the transport layer is to sort out what quality of service is needed – based on what is being provided by the network layer below – before the data is exchanged. This involves working out such matters as failure probability, throughput, likely transmission delay times, and error rates. In starting up a transport connection, the transport layer chooses the most cost-effective network service available consistent with the quality of service required. The transport layer keeps track of all the users of its services and what the limits are of its transport resources at any one time. Knowing this, it may offer a reduced quality of service for a particular user or, if a specified quality of service is unobtainable, may terminate the connection and inform the user.

One example of the way transport does this is multiplexing (Figure 4.2). The transport layer can decide to multiplex transport connections either upwards, from a single network connection on to several transport connections, or downwards, splitting a single transport connection among many network connections. Multiplexing upwards is cost-effective, since a number of transport connections make light use

The Seven Layer Model   65

Multiplexing upwards

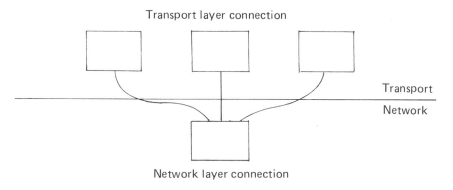

Multiplexing downwards

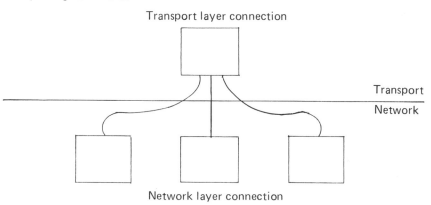

Figure 4.2 Transport layer multiplexing

of the network connections on to which they are mapped – this is explained further in section 4.6 on the network layer. Thus a number of light transport-users may make use of a single network connection. Conversely, higher throughput is possible if the multiplexing is downward, meaning that a number of network connections are made available to a single, downward-multiplexed transport connection.

The transport layer provides the session layer with a set of transport service access points (TSAPs) which give the session layer access to the transport service; and the addresses of these points. A transport connection is set up between two TSAPs, one in each of the connected systems, which are identified by their transport addresses. The transport layer must have the address of the local TSAP and the remote TSAP to which it will be connected. All TSAPs everywhere must have a unique address. There are many thousands of permutations and combinations of addresses. The transport layer has thousands of possible two-way simultaneous connections available to it between pairs of transport users. The transport layer knows what connections are available from the network layer below and manipulates them in order to provide the session layer above it with an end to end connection for a particular pair of systems. The systems will be unaware that their link is being maintained through the use of changing network resources.

When the connection has been established and the data is being transferred, the transport layer is responsible for preventing congestion on the network by flow control, the managing of a connection so that the transmitting system does not send the receiver more data than it can cope with. In MAP and TOP flow control is managed by queuing at source, which means the receiving entity – the active part of the layer – has to signal that it is ready for data before the data is sent. If the flow-control algorithm threatens to limit the throughput of data, the transport layer may overcome this by downward-multiplexing connections between the transport layer and the network layer below it.

4.5.2 Error correction
If data has to travel any distance at all, say from one factory or office site to another in another town – and especially if the town is in another country – the data will cross a number of different networks, some of which will be very reliable, others of which will be much less so. A LAN may be connected to a wide area network (WAN), say through a packet-switched network of the CCITT-recommended X.25 type, then to another LAN somewhere on the other side of the world. This transmission involves many networks, some reliable, some unreliable. The collective effect of transmission over a number of networks may mean that the the individual packets of which the data is made up may be duplicated or arrive in the wrong order. In more normal circumstances – office environments, for example – networks may be so reliable that the amount of error detection and recovery the transport layer is called upon to do is very little. Even there, however, the data might have been transmitted from a distance over a number of networks which are not so reliable and, as mentioned above, this may result in the data being in the wrong order and so on.

The local area networks in a manufacturing environment may be highly reliable left to themselves but the environment itself makes them unreliable. They will be subjected to noise and electrical interference which may corrupt the data. This and the fact that the networks are connectionless means the chance of transmitting corrupted data in a factory is high.

So the errors which can occur in a network are broadly of two kinds: those that are caused by the transmission of faulty data, and those where the data is correct but is corrupted during transmission. If the network service provided by the layer below is bad, the packets of data may be in the wrong order, they may be duplicated, they may be missing or they may be corrupted.

Something is needed to smooth out any deficiencies and check for errors. The transport layer does this. It restores connection-oriented services to the three applications-oriented layers above and carries out all the necessary error correction. The transport layer compensates for any deficiencies in the network layer and brings the service provided to layers five to seven up to a basic acceptable minimum. The transport layer is the great equalizer: it provides reliable, data transmission between two entities at the far ends of any group of networks – end to end, not just from one node to the next.

The transport layer uses a sequence number to check the sequence of the eight-number 'octets' of data and, if they don't come in the right sequence, it will resequence them or it will check that any particular octets are missing. It will check for duplication and it will recover. It will detect a lost octet and it will request a retransmission. To do this the transport layer may use end-to-end acknowledgement of messages, which means that every time a transport element receives a message it returns an acknowledgement of this to the sender. The sender transport entity retains a copy until the acknowledgement comes back. If it does not come back within a certain number of frames – called the 'credit', the sender transport entity will retransmit it.

This may suggest that, if it's wrong, transport will put it right. But this is true only up to a point. If an end-system – a system which holds a transport layer – crashes, the transport layer may not be able to recover lost messages. It cannot be assumed that all the messages that system sent out have been received, even if they were acknowledged.

Neither can it be assumed that unacknowledged messages were not received. This is because the transport entity at the receiving end has two jobs to do: pass the message to the correct session entity; and send the acknowledgement of receipt back to the sender. It cannot do both at once, and it may be that it sends the acknowledgement before crashing, and crashes before it sends the message to the session entity. Conversely, it may crash between passing the message up to session and sending the acknowledgement. This problem has to be sorted out either by the application or by the higher layers.

As described above, in manufacturing the network is operating connectionless and with a high risk of corruption or even network breakdown. This is so whether the network uses CSMA/CD or token passing. So the transport layer needs to pull out all the stops. In other cases the risk of corruption is negligible. This is why the service provided by the transport layer is divided into five types of protocol, depending on how much error recovery is needed. Which of the five will guarantee full duplex (two-way) transfer between two systems is negotiated during the call-establishment phase of the communication.

The five types are Class 0 to Class 4. Class 0 is the simple class which provides little error recovery at all. A Class 0 service will get and release the network, accept or refuse transport connections, transfer data and handle protocol errors. Class 0 is used where some errors are tolerable, as in teletex, for example. Class 2 is almost as basic but does provide multiplexing, explained above. These are only used where the network is reliable or where there are few resets and disconnects. Class 1 and Class 3 are used in networks which are reliable but where there is a high rate of resets or disconnects. Class 1 provides recovery from errors or disconnects signalled by the network, so it is is used where the network can be relied on to signal errors. Class 3 provides Class 1 error recovery and Class 2 multiplexing. It is used where the network reports errors but does not correct them. Class 4, the highest class, is used where the network cannot be relied on to report errors. It detects unsignalled errors and puts data in the right order if necessary. Both TOP and MAP specify Class 4 transport service. For manufacturing environments this is understandable since the environment is likely to be very hostile to any communications system. Only Class 4 recovers errors which arise from the corruption of data as it travels over the network – non-signalled errors, in other words.

One point to bear in mind about the use of OSI standards with local area networks is that the whole OSI model was constructed on the assumption that communications would be connection-oriented rather than connectionless. This has meant that the standards which allow connectionless working have been developed much later than those for connection-oriented working. However, the European Computer Manufacturers' Association (ECMA) has developed an architecture which uses a connectionless network layer to support the connection-mode transport layer. This is now being implemented in MAP and TOP systems.

There is also a CEN/Cenelec ENV41 102 preliminary functional standard which shows how to provide a connection-mode transport service between certain end systems – systems with a full seven layer stack – connected to CSMA/CD local area networks.

4.6 Network layer (layer 3)

Layer three, the network layer, allows data to be exchanged by transport entities using network connections. The network layer is not needed if communications are confined to a single network and need not be established across a number of networks. The network layer allows the transport layer to carry on without having to worry about which network a particular transport service access point (TSAP) is located, or about the routing needed to get to it, or about switching or interconnections, though network will tell transport what it has arranged about these. Network is responsible for providing an end-to-end path for the data, whether it is needed on the same network or on a distant sub-network. Network does the routing, switching and interconnection over sub-networks which may be public or private data networks or local area networks, and they may be token-ring, CSMA/CD, copper or fibre-optic. The network layer also notifies the transport layer of errors. In both MAP and TOP, the network layer must provide the connectionless network service (CLNS). This is specified in ISO DIS 8348.

Transport entities (see above) are identified by network addresses and the transport layer must map between transport addresses and network addresses. This is done by deriving network addresses from the transport addresses provided by the session layer. How this is done varies with the addressing scheme in use. MAP and TOP use hierarchical addressing, which means that, as the address in a packet of data goes down through the layers, the address relevant to that layer is removed before it goes down into the next layer. This means there is no need to store a lookup table to see which location corresponds with which address at each stage.

The network layer uses the unique station addresses which are passed down from the transport layer and determines a route through to the given address. To do this it needs to know the addresses of intermediate stations along the route, and it can get this information either from a look up table or from a network router station.

The network layer is really a number of sub-layers which are fully described in the TOP specification as follows:

The routing and relaying sub-layer
The sub-network independent convergence facility (SNICF)

The sub-network dependent convergence facility (SNDCF)
The sub-network access facility (SNACF)

In order, these are sub-networks 3.4, 3.3, 3.2 and 3.1 respectively. The routing and relaying sub-layer interfaces to the transport layer and is also called the 'Internet' sub-layer: it contains the routing information for data exchange from one end system to another end system, where an end system is defined as a node having a full seven layer stack. It supports an Internet protocol which works out how communication can happen across separate dissimilar sub-networks. The preferred protocol for MAP and TOP is the connectionless network protocol (ISO8473).

Below Internet is the SNICF, one function of which is to an interface between a connection-oriented and a connectionless network layer. It provides what the TOP spec calls a 'set of assumptions' based on the services available. The assumptions can include:

The expected shape of the network
The technical advantages and drawbacks of assuming one type of service rather than another
The likelihood that the assumed service will be available
The allowable complexity of the service

The SNDCF makes the service provided over a variety of sub-networks consistent. This can be done by enhancing some of the poorest sub-networks or levelling them all down to the lowest common denomimator.

The SNACF links directly to the datalink layer, layer two of the seven layer model. It is only needed where services must be adjusted as they cross the boundaries between networks. Indeed, as far as linked MAP and TOP systems are concerned, only the internet sub-layer will be used, which is why in MAP and TOP circles the descriptions 'network' and 'internet' are used, perhaps misleadingly, interchangeably.

4.7 Datalink layer (layer 2)

The datalink layer is not concerned with end-to-end communication; it is concerned with communication from one node to the next. It makes sure that frames are transferred error free from the network layer above it to the first intermediate point – on the same network – along the message's route. This point might be a bridge or a router's address and subsequent intermediate addresses are provided by the network layer. The datalink layer organizes media access to a single local area network.

The datalink layer resolves contention, does network re-tries when needed, detects noise and selects incoming frames which are addressed to its station. The datalink establishes connections upon request by the network layer and disconnects them in the same way. It corrects what errors it can and passes those it cannot correct up to the network layer. The datalink layer also supports some flow control functions.

The operation of the datalink layer is independent of the particular network-access method used – token bus, CSMA/CD and so on – used in the physical layer below it, and the access method is independent of the operation of the datalink layer.

There are many datalink protocols in existence. These protocols may be binary-oriented or character-oriented. Character-oriented protocols use control characters which vary from one supplier to the next. These could be a source of confusion if a control character used for one purpose in one system turned out to have a different meaning in the next, such as 'message begins' and 'message ends'. In binary-oriented systems certain strings of binary 1s and 0s define these functions without this ambiguity. A typical example used in wide-area networking is the high-level datalink control (HDLC).

In TOP and MAP one preferred choice in the datalink layer is logical link control (LLC). LLC is also called IEEE 802.2. There are three types of LLC:

LLC 1 is connectionless oriented
LLC 2 is connection-oriented
LLC 3 is single frame.

There are two classes of LLC service:

Class 1 supports LLC 1 only
Class 2 supports all three LLC types

Both MAP and TOP support Class 1, Type 1 but there is a special requirement in MAP to support LLC 3. This is discussed in the section on the enhanced performance architecture.

4.8 Physical layer (layer 1)

In the ISO model the bottom layer is the physical layer. Some ambiguity exists in the MAP/TOP community about the interface between the physical and datalink layer. This ambiguity revolves around whether the layer below LLC is a sub-layer of layer two or a sub-layer of layer one. But, since this is a matter of definition, it has no practical effect. Wherever it is, the sub-layer below LLC is medium access control (MAC). The MAC for MAP is the token-passing bus and the MAC for TOP is CSMA/CD, as explained elsewhere.

At the bottom of the physical layer is the modem, the broadband, carrier band or other modulator/demodulator which puts the signals on the network. These too have been described already. Below that is the cable itself. The physical layer makes and breaks connection with the cable medium on the instructions of the datalink layer. The physical layer also transmits news of any faults to the datalink layer.

4.8.1 Frame lengths
As noted above, the data which has to be transmitted for a typical application is added to as it goes down the seven layer stack. This means the physical medium has to send a lot more data to the destination than the mere message itself. And, even after all the layer control headers are added, the data containing message and its layer-specific information has to be assembled into a frame. A typical frame consists of:

> A preamble (one or more eight-bit sequences – octets) which set the clock of the receiving station so that the binary digits can be understood at the other end.
> A start-delimiter (one octet)
> A frame control signal (one octet)
> A destination address (two or six octets)
> A source address (two or six octets)
> The data (up to 8,191 octets, including control headers)
> A frame-check sequence (four octets), and
> An end-delimiter (one octet)

Apart from the data, then, there can be 19 or more octets to transmit. For long data strings, 19 octets takes comparatively little extra time to transmit but, if the messages are short, the 'overhead' becomes a respectable proportion of the whole transmission. Using the full seven layer model can add considerably to this overhead: some exchanges involve sending no data at all, as when two transport layers are merely exchanging acknowledgements. Another consideration is that 'overhead' data can take up several milliseconds of computation time at either end.

The time it takes to accomplish a complete transmission is lengthened considerably the more separate frames of data are transmitted. One function of the transport layer is to prevent the transmission of very short messages. The transport layer groups several such messages together into one longer message. If messages are short and frequent, the extra time taken to transmit all the extra signals –

apart from the message itself – can delay the message by many multiples of its own length.

This is particularly true of the messages encountered in real time control of machines or process plant, where messages are short, frequent and unpredictable. In such applications, transmission delays would be an important consideration under any circumstance but they are particularly important when something goes wrong. This is why the process control industry, whose members include all the large chemical, oil and gas companies, set about devising a fast communications system for real-time control many years ago (see carrier band, below). If a fire breaks out somewhere in an oil refinery the quicker and the more reliably the data must get through to spread the alarm. The same applies if a control malfunction causes a robot to run amuck. These conditions give rise to very short messages: in the case of the robot, there is really no need to expend a lot of time passing the robot information about why it must stop doing what it is doing. All that is needed is a 'Stop!' command which it should be possible to convey in a very short message indeed.

The special needs of real-time control applications have given rise to various proposals for speeding up the throughput of data in special circumstances and making the network more secure for sub-networks carrying this type of data. There are two main considerations. One is to cure the danger, slight though it is, that a head-end failure might bring down the whole network, even in an alarm state. This in turn implies using a single-frequency system – the so-called carrier band system. The other consideration is that the amount of overhead which such messages have to carry must be reduced.

4.8.2 Carrier band

Although constant reference is made to 'carrier band' as though it were a single, clear cut and readily available communications system, more than one carrier band system has been proposed. For some time before General Motors came on the scene, the continuous-process industries, represented by the Instrument Society of America (ISA), were working on a fast, real-time digital-bus communications system that could provide the fast response they needed. The process industries called their ideal system 'Proway'. It would use token passing and the high level datalink control (HDLC) data format already in use at the data-link layer in long-distance telecommunications. This specification was published in draft form as Proway A, later made more practical as Proway B.

This activity led to the formation of the US IEEE's 802 committee.

Originally the 802 committee was supposed to come up with a single communications system that would meet all industrial needs. But the initiative began to fragment under the pressure of the conflicting needs of different communications applications. Some needed a very fast system that would carry messages over short distances, others a system that would carry data for longer distances and so on. This fostered a proliferation of suggested access control mechanisms, addressing options, modulation methods, and speed and cabling options. It led to the division of 802 into 802.1 (station management functions), 802.2 (logical link control datalink sublayer), 802.3 (CSMA/CD), 802.4 (token bus), 802.5 (token ring) and 802.7 (slotted ring). All nine 802 committees are listed in Chapter 5.

There were even two different 'carrier band' options written into the IEEE token passing bus specification, IEEE 802.4. Including Proway, this meant there were three available carrier band options. In carrier band all the nodes on the network transmit and receive at the same frequency. Using a carrier gives higher noise immunity than is possible in a no-carrier baseband system. But since carrier band uses a single carrier frequency instead of the several used in a broadband system, there is no need to build a multi-frequency modem into each node on the network. Signals are impressed on the carrier by shifting the carrier frequency up or down. All the nodes are bidirectional: they can receive from or transmit to any other node. So there is no need for a head end.

All three proposed carrier-band systems used token passing. Token passing adds roughly 20 per cent to the overall transmission time but this is acceptable because it is predictable. All three also use frequency shift keying (FSK), a method of sending data using two signalling frequencies. But the type of FSK used varied between phase coherent and phase continuous FSK. And the method of coding differed among the three. In direct encoding two cycles of the higher frequency are used to represent a logic zero and one cycle of the lower frequency represents a one. In Manchester coding, one and zero are not represented directly by two different signal states but by a change of state from a particular signal condition to another. Thus 'high to low' might be equivalent to a logical zero and 'low to high' might be equivalent to logical one. If two consecutive zeros or ones have to be transmitted a transition state between the two from low back to high or from high back to low is needed, so Manchester encoding, which is used in Ethernet (IEEE 802.3), transmits data at half the signalling rate.

One of the two 802.4 carrier band options was very like Proway B. It was a 1Mb/s, phase-continuous, frequency-shift keying system which used Manchester coding. But it used a different cabling configuration from Proway and the short drop cables and non-isolating taps it specified were unsuitable, perhaps even unsafe, for factories because of worries about noise immunity, data rate and bit error rate. Long drop cables carry the signals away from sources of interference. Even when the short-drop option was crossed off the list, however, there remained two carrier band choices which allowed the use of impedance-matching cable taps and long drops away from the main trunk cable. The other 802.4 option used phase coherent FSK instead of Proway's phase continuous FSK. It used direct encoding as opposed to Proway's slower Manchester encoding. And it also provided two sets of data speeds – 5Mbit/s and 10Mbit/s – depending on whether logical zero and one were represented by 10MHz and 5MHz respectively or 20MHz and 10MHz. Proway uses 6.25MHz for logic high and 3.75MHz for logic low. Using Manchester encoding gives Proway a data transmission rate of 1Mbit/s. The losses per node for IEEE 802.4 carrier band are such that it can support up to 157 nodes where Proway cannot support more than 82 nodes.

When General Motors stepped in and identified the broadband option in IEEE 802.4 as the communications system which best met its needs, GM had no thought of adopting any of the available carrier band systems – even those in IEEE 802.4 – as part of MAP. For the reasons mentioned above, however, broadband token bus did not necessarily meet the needs of everyone else. The Proway committee had accommodated IEEE 802 by devising Proway C, an adaptation of Proway to make it consistent with 802 standards. But, where GM said its intention was to specify a broadband token passing bus working at a speed of 10MHz, the Proway people still felt sure a single-channel system working at 1MHz was the one that would best suit them. The Proway committee told GM that broadband solutions were much too expensive for their applications. They were also unhappy about the head end used in broadband systems.

The process control industries represented a constituency that GM felt it must have aboard the MAP bandwagon. And the process industries did not want to saddle themselves with a specification that needed special, therefore expensive, chip sets. Some compromise was needed between the two camps. Therefore GM and the, by now, growing MAP community accepted the need for a version of MAP which could be used for real-time applications.

In November 1984 an IEEE task group was established to study the 802.4 phase coherent carrier band specification and recommend any changes. Part of this involved setting up a GM-sponsored carrier band test-bed managed by Eastman-Kodak. After this happened a number of companies came in to compare phase coherent and phase continuous equipment. Phase coherent won over phase continuous and is therefore the preferred FSK method for MAP systems. The draft of MAP 2.2 published in July 1985 made provision for the first time for a 'MAP/Proway system' capable of use in either an OSI-based system or a Proway system. In effect the IEEE 802.4 phase coherent carrier band, though still the preferred carrier band version, it has been modified to make MAP and Proway look more alike. Differences between Proway and the 802.4 carrier band version are being eliminated on the Proway side too, however. Late in 1986 Proway C's phase continuous FSK modulation was voted down in favour of phase coherent FSK. It is likely that Proway C – now IEC standard 955 – will eventually use higher data rates. This is of benefit because a closer match to the 10Mb/s data rate of the main MAP broadband backbone enables easier links with carrier band sub-networks. It will also give Proway users the benefit of the cheap, single-chip carrier band interfaces now being developed for MAP networks.

4.8.3 Enhanced Performance Architecture (EPA)
No matter what modulation and encoding techniques are used in real time networks, the elimination of head ends only deals with one issue – security – among many for users of these networks. There remains the need to reduce the amount of data which has to be transmitted in order to achieve the sending of short messages at frequent intervals without too much data overhead. The main method so far proposed is to dispense with some of the layers which add data to the message as it filters down through the OSI model. One way of doing this would be to provide 'hooks' at the LLC (logical link control) in layer two which would allow the application to communicate without having to go through the top five layers of the seven layer model. A system which provides all seven layers but allows access at the top of layer two is called the enhanced performance architecture (EPA) (Figure 4.3). Some systems could be provided with only the bottom two layers of the seven layer stack. This two layer version of the full MAP stack is called mini-MAP (Figure 4.4).

Another slight but important change would speed up some data exchanges. In a normal token passing network, a node which receives a message cannot reply to it until it gets the token. The node it replies to cannot send a further message in response to the reply until it gets the token a second time. LLC type three would be used in the EPA because it allows short exchanges of data between nodes on a network when

The Seven Layer Model 77

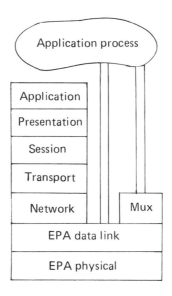

Figure 4.3 MAP/EPA architecture

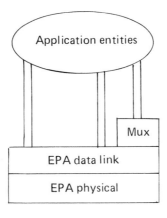

Figure 4.4 Mini-MAP nodes

either of them holds the token. This would be an obvious advantage in a real-time application. It allows quick question and answer sessions about important matters like 'Is the chuck open or shut?'

But some in the MAP and TOP community think it of little help to eliminate the layer-dependent data overhead when the more significant source of data overhead is the framing data. For this reason it has been suggested that the EPA should exist in a version which allows nodes on the network to use the services provided in all seven layers but to use them operating in connectionless mode. This would mean fewer data-less protocol exchanges between between transmitting node and recipient node and reduce the number of frames of data which would have to be constructed. But this solution leaves open once again the question of security on that part of the network. If the network failed, it would be some time before it was realized that some item of time-critical data had not reached its destination.

4.8.4 Why bother with broadband?

Whatever form the EPA takes it will run on a carrier band sub-network. It is often said, especially in Europe, that many are likely to find carrier band a more attractive alternative to full broadband as a basic factory-communications network. However, others who have studied MAP systems at close quarters are beginning to believe that broadband, on balance, has worthwhile advantages over carrier band and EPA.

This is worth examining in more detail. Concord Communications of the USA, for example, supplies both carrier band and broadband systems but finds that most of its sales are in broadband systems. CCI president Tony Helies admits that the take-up of broadband varies from place to place, within areas of Europe and even within areas of the USA. 'A number of companies are doing broadband projects, and they have been driven to do them by the existence of MAP. And after all you can do so much on your broadband. Carrier band needs a similar cable – it doesn't look all that different from broadband cable – and yet just one channel absorbs all the cable. You can only put that one channel on it.'

Roy Cadwallader of ICL of the UK agrees: 'Get the cable in first, that's the first step. Putting in the cable is the biggest disturbance, whichever type of cable it is. And the disturbance is the most expensive part of it.'

ICL is currently in the early stages of an ambitious MAP broadband project (see Section 7.7). ICL's Jim Kenny is deeply involved in the project and his view is that broadband will enable companies to do so much more with their networks that they will do well to consider it right from from the start. ICL plans some novel uses for its broadband cable. For example, ICL plans to offer operators video information to prompt them through unfamiliar tasks. Kenny believes this will be such an effective training aid that the extra production obtained from quicker-trained operators will more than outweigh the cost of preparing the video material the operators will use. ICL also plans to run Ethernet-type communications on the same broadband cable as the MAP token-passing network.

Ferranti of the UK says broadband users will also gain from the increasing interest in the technology shown by banks, insurance companies, hospitals and airports. Ferranti, which installed the cable for the UK's CIMAP demonstration in Birmingham UK in 1986, says there is a lot of interest in broadband among institutions in the City of London. The systems the financial community is interested in would be based on IEEE 802.3 at the physical layer so this is likely to prove a big stimulus for the TOP. In airports there is a lot of interest in using broadband systems to carry TV pictures as well as baggage-handling and other computer data. And hospitals are using broadband techniques to move patient-data around. In Canada broadband systems are used to provide access points for patient-monitoring systems moved round on trolleys. And of course one main criterion of hospital systems is that they must never break down.

All in all, a broadband bus still looks attractive. Cheaper, proprietary networks can be attached to the bus through gateways, reducing the number of modems needed per shop floor device. This means the shape of the networks in companies would be more like a tree than a straightforward bus, with local networks in particular parts of a company transporting data among local functions, then linking into the main broadband cable. The local nets can even be proprietary systems, though it seems inevitable that even proprietary nets will take on more and more features of open-systems interconnection.

4.8.5 Field bus
There is one more communications type which MAP will probably have to accommodate: the field bus. Where carrier band links devices at plant floor level, there is a need to provide a system which links the sensors on the plant floor. It is much too expensive to provide every switch and sensor in a plant with even the simple single-frequency interfaces used by a carrier band network. At sensor level you need a simple, effective, cheap and very rugged means of wiring everything up to the programmable controller or other device that sits on the carrier band network. Such a system may also be required to connect

measurement and test systems and to provide internal links in machine tool and robot controllers. It has to deal with frequent, short messages.

The field bus is the solution. It will succeed the four to 20 milliamp current loop communications system. Current loop was devised by the SP50 committee of the Instrument Society of America. The ISA represents the process control industry in the USA. Current loop is used not only in process plants but to link switches, actuators and other rock bottom plant items on the factory floor.

In October 1986 SP50 met to prepare its final draft for a field bus standard to replace current loop. SP50 does not determine the final form of the field bus. That is up to working group six (WG6) of International Electrotechnical Commission committee TC65. WG6 drew up a detailed draft of the requirements a field bus standard must meet. But SP50 has great influence on what WG6 decides. SP50 and WG6 had a joint meeting after SP50's October 1986 meeting and it was only then that WG6 met in Phoenix to prepare its final draft.

What WG6 decided was what the field bus must be capable of doing. WG6 specified the requirements of two types of field bus: a fast bus communicating over 40 metres; and a slower bus communicating at up to 350 metres. Among the many field bus systems on offer already, the most popular short, fast bus must be the US Mil-standard 1553B, a system devised by the US Air Force to provide foolproof data exchange in military aircraft. It has been around for a decade or more and there is a lot of equipment on the market which meets it. The companies which supply it to the US military are anxious to widen their market by having its use extended to industry.

ERA Technology in the UK is part of a consortium which is developing a field bus system based on 1553B. ERA says it has put 1553 through severe testing, particularly of its susceptibility to electrical noise, and been unable to fault it. ERA and its collaborators – Allied Amphenol, the systems and integrated circuits divisions of Marconi Electronic Devices, the hybrid units of STC Components, the Taylor Instrument division of Combustion Engineering, Technitron International and Loral Instrumentation – are developing 1553 in two ways. One is to extend it to cover the slower, long distance requirements of WG6. The other is to make both versions of 1553 intrinsically safe and able to deliver power to the sensors that need it. Both these requirements, which tend to work against each other, may be beyond what WG6 will require.

But there are those who say a military system is unsuited to use in commercial and industrial communications. The chief objection is that the temperature requirements, for example, are too stringent to be relevant and make the cost of the system too high. This is why Intel, which is also very influential, has put forward its Bitbus. The French government is also backing a system called the factory instrumentation protocol (FIP). US electronics company Motorola has also made a late entry into the market.

In 1987 a West German consortium proposed still another field bus system. The West German consortium, which includes Siemens, Bosch and Klockner-Moeller, believes 1553 is not only too expensive but too slow.

Their solution is Profibus, a contraction of 'process field bus' It uses token access, which makes it compatible with MAP, and works at up to 500,000 bytes a second. The project began its definition phase in May 1987 and the consortium planned to try it on a production line in March 1988. Multi-vendor applications of the system were to begin at the end of 1988. Those involved put the cost per sensor at $50 to $100.

Standardization is unlikely before 1990 or 1991.

4.9 Interconnecting MAP and TOP with each other and with other systems

One of the most important features of the OSI seven layer model is that it allows the substitution of different protocols – which should also be standard protocols – for each of the levels in the seven layer stack. But since the main purpose of open systems is that OSI-conforming systems should be able to communicate with each other, this implies that an implementation which uses one seven layer set of protocols should be able to communicate with another implementation which uses a slightly different, or even completely different, set of protocols.

Such translations are accomplished by 'gateways' (Figure 4.5). The gateway has two architectures, one for each set of seven layer protocols, and it transfers information peer-to-peer between them just as communicating peer-level entities at either end of a network do. The data climbs to the top of one seven layer stack in its original form and goes back down the second stack, gathering protocol control information which conforms to the protocols adopted in the second stack.

Gateways are necessary for translations between, say, a MAP network and a wide-area network such as a CCITT-approved X.25-type telecommunications network – such gateways were demonstrated at the UK's CIMAP communications demonstration. Gateways also translate data flowing in one type of LAN to protocols which allow it to flow on a different type of LAN. This means that gateways will be a key method

82 MAP and TOP: Advanced Manufacturing Communications

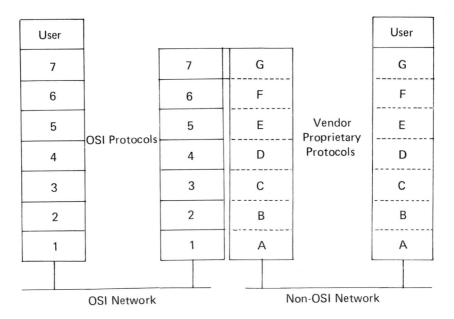

Figure 4.5 Gateway architecture

of migration from non-MAP and non-TOP to MAP and TOP systems: it should be possible to interconnect any proprietary LAN, such as those provided by the programmable controller suppliers, to a MAP or TOP LAN through a gateway.

Gateways are complex, and therefore expensive. And the same features of the full seven layer stack that led to the promotion of the EPA also apply to gateways. This means that they are slow to operate because peer-to-peer end-to-end negotiations are only possible if every transaction is translated through the gateway. In fact General Motors regards the use of gateways as unacceptable and they will be phased out after a time.

There are simpler ways of interfacing certain types of OSI-conforming systems than using a gateway. One is the router and the other is the bridge. Much of the MAP literature confuses the two, and the generally used definitions here, though conforming with the TOP 3.0 draft specification, conflicted with those used in the MAP 2.2 draft published by the Society of Manufacturing Engineers for the US MAP Users' Group.

The 'bridge' connects any two compatible networks so effectively

The Seven Layer Model   83

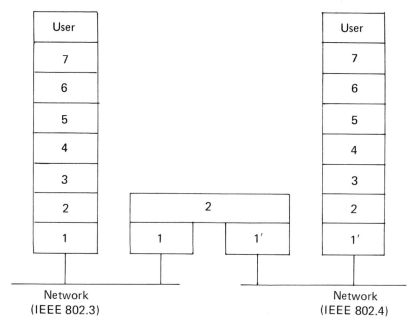

Figure 4.6 Bridge architecture

that the transmitting node need not even be aware of the bridge's existence (see Fig. 4.6). Bridges have memory buffers which can store information for later transmission. A bridge might connect two MAP systems operating in different broadband channels or a MAP broadband backbone with a carrier band sub-network. Bridges can be used to isolate two similar LAN sections. Bridges make their links at the datalink layer but the two networks linked in this way must use the same addressing scheme and frame size.

This is not necessary for the 'router', also referred to as an intermediate system architecture. The router interconnects two or more different networks with a common network protocol which routes data through from one three-layer stack to the one relevant to its destination (see Fig. 4.7). In the router, which the specifications also call the 'intermediate open system' or 'intermediate system', the data has to travel one layer higher, to the network layer, because, as already explained, only the network layer can guide traffic across a number of subnetworks. The router is given a distinct network address which all the networks attached to it are aware of. The connected networks may

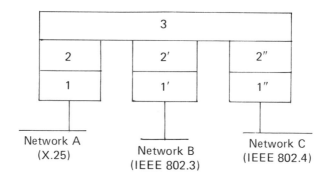

Figure 4.7 Intermediate system architecture (router)

use any datalink (MAC) or physical layer protocols, so routers can be used to link different MAP systems or MAP and TOP systems, though the simpler bridge (see above) can also accomplish this provided the MAP systems are wholly interoperable. If not, routers can connect non-interworking MAP systems together. For example, as explained elsewhere, early interfaces supplied by Industrial Networking needed to use INI head ends, and Concord interfaces needed to work with Concord head ends. As IBM and DEC demonstrated at CIMAP, a router enables data which uses an INI interface to get on to the MAP network to reach a node connected to the MAP network via a Concord interface. An Intel router carried information between INI and Concord interfaces on the main MAP backbone and between them and the TOP network that was also running.

A 'repeater' (Fig. 4.8) extends the length of a LAN by carrying signals from one LAN segment into another LAN segment with exactly the same physical characteristics: that is to say, the link is achieved at the lowest, physical layer. This means, for example, that a token generated in one LAN may cross the repeater into the next LAN. Repeaters can alter the transmission medium, so that data may be passed from coaxial to fibre optic cable, for example.

4.10 Technical and office protocols (TOP)

As mentioned earlier the difference between MAP and TOP lies in the environment and application in which they will be used. The Technical and Office Protocols, as the name implies, addresses the problem of providing multivendor compatibility and communications in the technical and office functions. TOP is the initiative of a group of major computer systems purchasers. The members of TOP include Boeing,

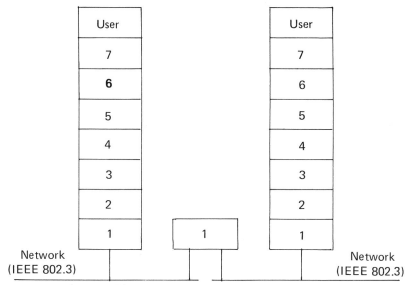

Figure 4.8 Repeater system architecture

General Motors, Du Pont, McDonnell Douglas, Ford Motors, Procter and Gamble and Eastman Kodak.

These major companies are involved because they consider that the rapid achievement of open systems interworking using international standards will provide them with the basis for much improved communications, both internally and externally, and make it much easier to incorporate innovative products into their computer systems.

Boeing Network Architecture (BNA) was developed in 1979 to address the multi-vendor interconnectivity problem within the Boeing company. BNA defines a three-tier model for computing and data communications. Tier 1 contains the data centres and corporate servers, which generally are characterized as mainframe computers. Access to corporate wide data bases is at the Tier 1 level. Tier 2 contains divisional and working group servers, generally characterized as minicomputers and super-minicomputers. Tier 3 is the workstation, generally characterized as microcomputers and super-microcomputers.

In a multi-vendor computing environment, common communication methods are required for integration and automation of processes. The Boeing company's stated direction for all corporate data communications is:

> The ISO/OSI Reference Model and OSI conforming protocols are identified as the long range BNA direction for the direct interchange of information between different vendors' equipment.

Since TOP is an application oriented implementation of the OSI protocols, it conforms to the BNA stated direction. The core suite of protocols in TOP are the OSI protocols with option and parameter selection for the protocols reflecting the choices made at the NBS OSI implementers' workshops and published in the NBS *Implementation Agreements Among Implementers of OSI*. The functional extensions of TOP are intended to follow the ISO standards. These extensions, therefore, also conform to the stated direction.

According to the OSI reference model document [ISO 7498]:

> The purpose of the International Standard Reference Model for Open Systems Interconnection is to provide a common basis for the co-ordination of standards development for the purpose of systems interconnection, while allowing existing standards to be placed into perspective within the overall Reference Model.

The ability to use the OSI model to encompass existing and emerging standards for data communications and migration purposes is one of the major reasons for choosing it for the TOP network architecture.

So TOP, like MAP, is based on the seven layer OSI model. But, unlike MAP, TOP does not have to address the needs of a harsh environment. And the applications which it is designed to integrate are different: electronic mail, word processing, document exchange, file transfer, graphics interchange, database management, batch processing, videotext and business analysis. Some of these functions are described in section 4.2.5. TOP will provide peer-level, process-to-process, and terminal-access communications for computers ranging from microcomputer workstations to mainframes. TOP will also interconnect office tasks with the factory floor via its interface with MAP. TOP and MAP are intended to use the same base suite of protocols. These common protocols of the initial TOP specification are:

Layer 7 – ISO File Transfer, Access, and Management (FTAM) [ISO 8571/1-4]
Layer 5 – ISO Session [ISO 8327]
Layer 4 – ISO Transport [ISO 8073]
Layer 3 – ISO Connectionless Internet [ISO 8473]
Layer 2 – IEEE 802.2 Logical Link Control (LLC) [ISO 8802/2]

The only areas in which TOP and MAP initially differ are:

Layer 7 – Manufacturing Message Service (MMS) not required by TOP.
Layer 1 – IEEE 802.3 CSMA/CD Media Access Control (MAC) [ISO 8802/3]

The Seven Layer Model 87

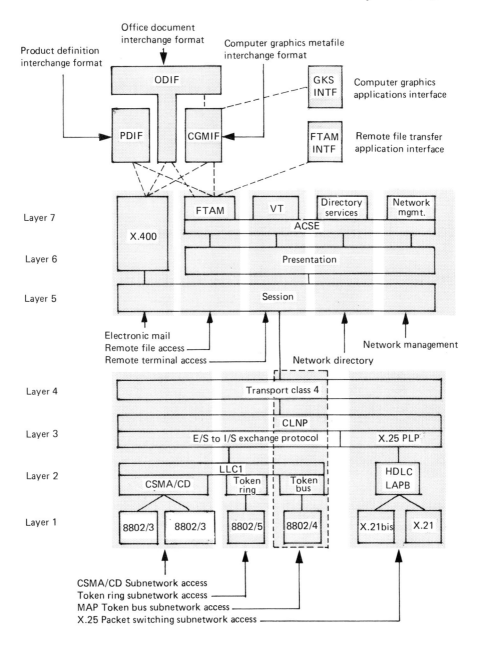

Figure 4.9 Building block overview (TOP 3.0)

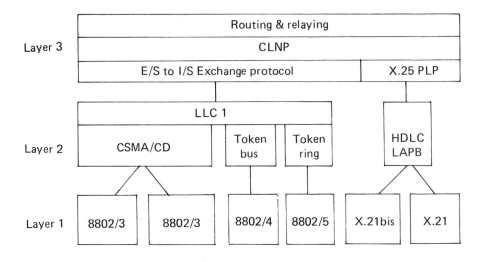

Figure 4.10 Intermediate system building block (TOP 3.0)

TOP builds on and complements the work done by MAP. In many cases the protocols selected are the same and there is an intention to avoid any needless differences.

TOP 3.0 includes three different transmission protocols. As well as baseband 802.3, it includes broadband 802.3 and token ring, 802.5.

TOP will accomplish several major architectural objectives, such as allowing for the interconnection of multiple office LANs and for connection to WANs and digital PBXs for long distance communications.

In the latest published version of the TOP specification (3.0), various options are described for various parts of the TOP seven-layer model. These are shown in Figures 4.9 and 4.10, which are fairly self-explanatory. Dashed lines show potential binding associations between the upper-layer building blocks and the interchange format and application interface building blocks. According to TOP 3.0, a complete TOP end system will include one or more building blocks covering layers one to four and at least one building block for layers five to seven.

The operational benefits expected from the above would be a standard user office communication network, allowing easy multi-vendor data access and data interchange, lower office systems costs would come from reducing the need for multiple cables and customized networking software. Users would be given more choice of suppliers based on communmication compatibility between different systems. Flexibility and adaptability of production systems would improve to meet changing demands and shorten lead time in designing and implementing integrated office systems.

As noted in section 4.2.4.2, TOP should build on the office document architecture (ODA). This work has been undertaken by the European Computer Manufacturers' Association (ECMA) and is currently approved as draft international standard, DIS 8613. ODA is the key component of the set of open systems standards for office systems, as it defines the structure and transfer format of compound multi-medium, multi-format documents.

As already explained, the TOP electronic mail and messaging standards are based on the CCITT X.400 standard, which was ratified in 1984. This will probably be the first full seven layer OSI 'stack' to be ratified.

5
THE STANDARDS SCENE

5.1 The origins of OSI

It is important for both vendors and users to understand the process by which international standards are derived as well as the original concepts and objectives of Open Systems Interconnection (OSI). This is because the standardization work on OSI is different to traditional standardization in at least one key respect: it is prospective. This means the standardization happens before products reach the market place. It also means that standards development is effectively an integral part of product development. The standards, as they are developing, therefore point the way in which the technology is moving. This cannot just happen by accident: it needs the prior commitment of all the major manufacturers to making sure that the eventual products they produce will meet the standards.

For the users of the resultant kit, knowledge of the relevant standards confers knowledge of the functionality that will be provided by future systems. This knowledge allows them to plan their procurement strategies with confidence; anyone familiar with the anarchy of the computing market place will immediately recognize the major achievement of the International Standards Organization (ISO) in this area of its work. Those who wish to accelerate, influence and access the standards must study the standards process to understand how they are developed.

OSI developed because, as the major computer manufacturers started to develop their own communication architectures, those manufacturers realized that interworking between these architectures was going to be a major problem. They realized too that this problem would be expensive to resolve and would bring in its train two serious extra disadvantages:

> The incompatibility problems would discourage potential users and the size of the total communications market would be drastically reduced;

The users would inevitably gravitate toward the largest manufacturer, leaving other manufacturers with a major headache: do they persist with their own architecture or do they follow (at some distance) the market leader?

The users also realized the dangers of being locked into one supplier for their communications requirements. They lent their voices to demands for a vendor-independent architecture which would create a true multi-vendor environment, one in which they could purchase kit from different suppliers secure in the knowledge that every item of such kit would, if it conformed to the Open Systems standards, interwork with every other such item. As things turn out, it is not quite as simple as that, but tremendous strides have been made towards that goal.

The OSI vision was first conceived in the late 1970's and gathered momentum. Big organizations, including government departments in the USA and the European Community, publicly announced their support for the OSI initiative. Also important was the close collaboration between ISO and the Comité Consultatif International de Télégraphie et Téléphonie (CCITT), a sub-agency of the United Nations' International Telecommunication Union. This collaboration led to the publication of technically identical OSI standards by those two bodies. And General Motors' and Boeing's decision to base MAP and TOP on ISO was crucial both to the acceptance of OSI and the speed with which it developed.

Now, any vendor or user currently thinking of marketing or buying communications equipment has to make an assessment of that equipment's ability to interwork using OSI protocols; for OSI is moving rapidly from the drawing-board through the implementation phase into the market place.

5.2 The International Standards Organization (ISO)

MAP and TOP are each, in effect, functional implementations of OSI, covering all seven layers of the OSI Basic Reference Model. In fact, it is intended that the only standards that MAP 3.0 and the equivalent TOP specification will reference are the ISO standards for OSI and LANs. So the development of MAP and TOP is constrained by the availability of these OSI and LAN standards from ISO. In addition, the full range of OSI standards will be extended to include a range of functionality which is not directly required for the MAP 3.0 and TOP 3.0 specification but is essential for interworking between MAP networks and other functional implementations of OSI, such as TOP.

The ISO was the logical choice for the development of the OSI family of standards because it offers:

A non-denominational forum, that is, a forum dominated by neither vendors nor users;

Established procedures for international collaboration on the preparation of technical specifications;

The end product has the status of International Standard and so is accorded wide recognition.

The way ISO operates is to weld nationally formulated views into an international agreement on a particular draft text or technical issue. The cost of international travel and subsistence, not to mention the cost of the experts' time, means that, where possible, the ISO work is undertaken by correspondence. Meetings are held only to resolve national differences on specific issues and to decide co-ordination and policy issues.

The 'overhead' activities are undertaken by the staff of the ISO Central Secretariat in Geneva. They are responsible for the overall management of ISO Technical Committees (TCs), publication of standards and so on but they do not get involved in the technical issues; these are managed by the TCs themselves. The TCs in turn are administered by decentralized national secretariats, usually national standards organizations like the American National Standards Institute in the USA or the British Standards Institution in the UK.

The Central Secretariat is also responsible for editing the draft standards according to the ISO/IEC rules for the presentation of standards. The idea of this is to achieve a consistent format but, curiously, in OSI this is not a major task. This is because the OSI project editors (see below) work very closely to the ISO rules; because the OSI standards follow a well established and consistent pattern, for example service definitions and protocol specifications; but mainly because the subject matter is so complex that only experts in the subject can edit the drafts. All this means that although the TCs are working within ISO procedures, they are largely autonomous and can determine their own work programmes and rate of progress.

There are now nearly 200 ISO TCs. Each comprises members from those ISO member countries that wish to be involved in that area of work. The members may be either participating members or observer members. The TCs may also use a liaison category of co-opted members from other international organizations who have a legitimate interest in the work. The two ISO TCs relevant to the work on MAP

94 MAP and TOP: Advanced Manufacturing Communications

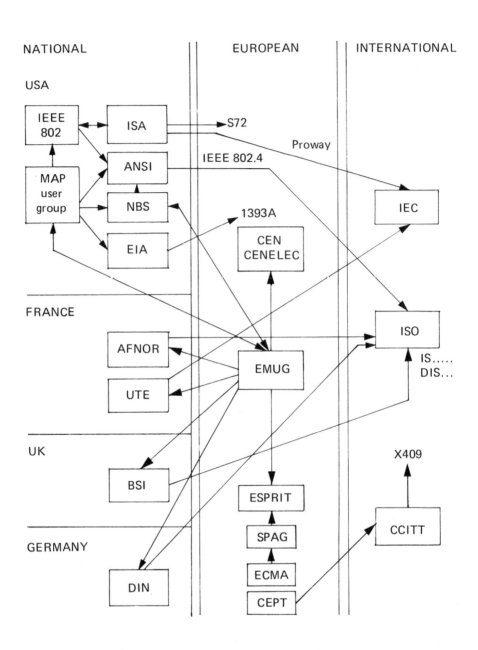

Figure 5.1 Standardization bodies

are TC97 'Information Processing' and TC184 'Industrial Automation'. TC97 is responsible for the OSI standards, and has a number of liaison organizations including CCITT and the European Computer Manufacturers Association (ECMA). It is expected that this liaison category will be extended to include the MAP/TOP user groups.

The real technical work is, however, undertaken not by the TCs but by their subcommittees (SCs). The two SCs of TC97 responsible for OSI are SC6 (which deals with LANs and layers one to four) and SC21 (which deals with layers five to seven, graphics and databases). SC4 of TC184 deals with the Initial Graphics Exchange Specification (IGES) and related matters. These SCs have their own membership and liaisons on a similar basis to the TCs, and each has a substructure of Working Groups (WG) which broadly correspond to the layers of the OSI Reference Model.

TC184 has recently become active in the area of inter-device communication in the manufacturing environment, and is currently responsible for work on the Manufacturing Message Service (MMS) which is being undertaken in conjunction with the Electronic Industries Association (EIA), the American organization responsible for the RS series of standards, so that the resultant standards (ISO 9506 and EIA RS511) will be identical.

The International Electrotechnical Commission (IEC) is involved at the fringes of this work, their most notable contribution being IEC 955 Proway C. This work was done by IEC TC65C (see section 4.8.2).

5.3 ISO stages of development of standards

There are several stages in the development of an ISO standard and the work programme is rigorously controlled to ensure that target dates are met. The stages are:

NWP – New Work Proposal

NWPs can be generated by a member country of the TC, a liaison organization or the TC or SC itself. NWPs are generally no great surprise, usually being called up from a rolling programme of work. If an NWP is to have any chance of success, it must be well-defined in scope and clear about the form of the eventual output: for example, a new ISO standard or an addendum to an existing standard. A successful NWP must also be justified, as it will be competing for resources with other projects. Thus ISO effects stringent control over its work programme. The NWP is then subjected to national ballot at the TC level and to succeed it needs not only a minimum number of member countries prepared to actively participate in the project.

WD – working draft
When an NWP has been accepted, it will be assigned to a particular SC and WG who will prepare an initial draft, the WD; this WD may be based on a draft submitted with the NWP or it may be prepared from scratch. The WD is considered unstable and is not generally circulated outside the WG and SC.

DP – Draft Proposal
When the WD has passed through several rounds of study and refinement, it will reach a level of stability and can be issued for a two month ballot and comment by the member countries of the SC as a DP. This ballot period is, in effect, a moratorium on the development of the draft while it is subjected to national study.

DIS – Draft International Standard
The comments on the DP will be studied in detail by the WG and efforts will be made to resolve any votes of disapproval without, it is hoped, causing any approval votes to change to disapproval. After this stage, the draft has reached a level of agreement about its technical stability: if not, it may be necessary to repeat the DP stage. Once approved, the draft can be issued for six-month ballot by the member countries of the TC as a DIS. It now becomes apparent why international standardization is a long process, although it should be noted that the draft will be considered suitable for issue as an 'intercept' (for example as a BSI Draft for Development) for the guidance of potential implementors and users.

IS – International Standard
When the text is published as an international standard, it is considered to have reached full stability. However, the process of standardization in these leading-edge technologies is dynamic, and even the ISO standard is subject to a process of continual review and updating, usually through the issuing of addenda.

One more point which should be explained because of its special significance in the MAP and TOP specifications is the relationship between the IEEE 802 committees and the relevant ISO committees. The IEEE 802 committee was originally formed to devise a standard LAN. Only later did it become apparent that several sub species of LAN would have to be accommodated. This led to the spawning of a number of 802 committees shown below. The IEEE submitted its LAN proposals to TC97 through ANSI, which for the purposes of IEEE 802, is the ISO member-body representing the USA.

802.1 is station and network management functions, called high-level interface functions and so on
802.2 is logical link control, datalink sublayer
802.3 is CSMA/CD
802.4 is token bus
802.5 is token ring
802.6 is metropolitan-area networks (MANs)
802.7 (see below)
802.8 is fibre-optic systems
802.9 is integrated voice and data networks

Many of these are not draft or draft-international standards, merely the numbers of committees which meet to discuss certain topics. But where they do refer to standards which are going through the process of becoming DISs, the ISO has reserved numbers for those eventual standards. The reserved ISO numbers have been chosen so that it will be easy for those affected to tell which standards are which: the ISO numbers merely add an extra '8' to the front of the IEEE number and replace the '.' with a '/'. Thus IEEE 802.4 becomes OSI DIS 8802/4.

In one case the numbers do not coincide. The 802.7 committee of IEEE is the broadband technology advisory group, but the UK submitted a proposal to ISO that it reserves the number DIS8802/7 for the slotted ring or so-called 'Cambridge ring'. The IEEE 802.7 group is carrying out work, however, which will not produce a standard, so there will be no conflict with 8802/7.

5.4 National activity in the UK

In the UK, the national activity is co-ordinated by the British Standards Institution (BSI) as the national standards organization. BSI has a strong structure of committees and expert groups which actively participate in the OSI standards work. In general, most work undertaken in BSI committees is directed toward inputting UK contributions into ISO and deciding on UK votes and comments on drafts issued for ballot, and BSI committee structure is in effect a mirror image of the ISO committee structure.

Early in 1987 BSI set up a new scheme, called BITS (BSI Information Technology Services), to provide rapid information on developments in the area of IT and AMT standards activities. Details of this service can be obtained from BSI, Linford Wood, Milton Keynes, MK14 6LE.

5.5 Who's who in standards

Within the United States there are over 400 organizations involved with setting and adopting standards and many others worldwide. A list of some of the most prominent US and international organizations follows.

American National Standards Institute (ANSI)

This group, based in New York City, is the US member of the International Organization for Standardization. It co-ordinates the writing of many US domestic standards and approves them.

Computer & Business Equipment Manufacturers Association (CBEMA)

This trade association of equipment manufacturers holds the Secretariat for the X3 Information Systems Committee.

Computer and Communications Industry Assocation (CCIA)

The CCIA is a leading US trade association of computer and communications equipment manufacturers.

Corporation for Open Systems (COS)

This consortium of vendors and users was formed in 1985 to accelerate the development and commercial availability of interoperable computer and communications equipment and services that conform to ISDN, the Open System Interconnection model, and related international standards. It now has more than 50 members.

Electronics Industries Association (EIA)

This trade association represents manufacturers of electronic equipment and components, telecommunications equipment, radios, and televisions.

European Standards Coordinating Committee – European Committee for Electrotechnical Standardization (CEN – Cenelec)

Major equipment manufacturers join European standard-setting bodies and user groups in this organization for Common Market members. It has the main standardization role for computers in Europe.

European Communities (EC)

This key European administrative and policy-making group – the Common Market – seeks harmonization of telecommunications standards. Among its motivations are the perceived lack of European innovation and competitiveness in the design and production of computer systems.

The Standards Scene 99

European Computer Manufacturers Associations (ECMA)
Despite its name, this Geneva-based technical group does more work in setting international standards than as a trade organization for its equipment makers. ECMA standards often become the basis for ISO and IEC information processing standards.

European Conference of Postal and Telecommunications Administrations (CEPT)
All European telecommunications carriers belong to this group based in Bern, Switzerland. It has the main technical role in the European Community's efforts to standardize telecommunications standards.

Information Technology Steering Committee (ITSTC)
This newly formed group, based in Brussels, Belgium, has the aim of co-ordinating European efforts on standard functions for the OSI model. It incorporates the work of CEPT and CEN-Cenelec.

Institute of Electrical and Electronics Engineers (IEEE)
In this role as the technical and professional association for electrical engineers, the IEEE recommends standards in many areas of electrotechnology. It acts internationally through the US National Committee for the IEC, and through ANSI (the US member-body) for ISO.

Instrument Society of America (ISA)
The ISA represents the US process-control industry. Its SP50 committee devised the 4-20 milliamp current loop and, because of its influence in the International Electrotechnical Commission, it has a strong say in the field bus standard which will replace the current loop.

International Consultative Committee for Telegraph and Telephone (CCITT)
The national telecommunications companies of the world gather in this organization, chartered by the United Nations, to set telecommunications standards. CCITT has a great deal of input into ISO standards, and final standards issued by the two groups in areas of common interest are frequently identical. The US State Department heads the US delegation to CCITT.

International Electrotechnical Commission (IEC)
Founded in 1906, this international standards organization has a membership comprising 43 countries that produce over 95 per cent of the world's electrical energy. It develops standards covering the entire field of electrotechnology, including electric power apparatus,

electronics, appliances, radio communications, and transportation equipment.

International Organization for Standardization (ISO)

This worldwide voluntary standards associations, based in Geneva, is made up of 77 countries' standards-setting bodies. It has over 150 technical centres, writing standards in many areas. One task, for example, is to create OSI standards through Subcommittee 21 of its Technical Committee 97 for Information and Data Processing.

International Telecommunications Union (ITU)

This is the international organization responsible for the adoption and assignment of international communications protocols and frequency allocations. It does its technical work through its two committees, the International Radio Consultative Committee (CCIR), and the CCITT.

Manufacturing Automation Protocol & Technical Office Protocols (MAP/TOP) Users Group

This group includes professionals and corporations involved with communications integration in the areas of engineering, office, and manufacturing automation. It became a technical group of the Society of Manufacturing Engineers (SME) in 1985. It works to develop existing and emerging international communications standards and accepted operating practices.

National Bureau of Standards (NBS)

The NBS is the US government organization responsible for basic physical and measurement standards in the United States, and for research and development of scientific measuring methods. It also develops computer and information systems standards used by the US government, called Federal Information Processing Standards.

National Electric Manufacturers Association (NEMA)

This old and influential trade association of electrical equipment manufacturers develops standards for products produced by its members.

Standard Promotion and Application Group (SPAG)

This trade group is composed of 12 European computer makers, under EC sponsorship, who provide technical recommendations to ITSTC.

6
KEY ISSUES

As will already be clear, neither MAP nor TOP is a finished set of specifications. There are years of work ahead in terms of deriving standards for some functions and settling the practical details of implementing even those parts of the specifications that are already stable. The point of this chapter is to raise just a few issues of immediate concern. For those who have to do all the work, issues of 'immediate concern' are a movable feast – the issues that are of concern today will be quickly replaced by other problems as time elapses. For the moment, however, the points that seem to be prompting the biggest questions are migration, testing, and the possible unforeseen consequences of standardization. They are placed in this part of the book because the reader will by now be familiar with most of the matters that bear on these issues. First we will deal with standardization and its possible effects on the manufacturing messaging service (MMS).

6.1 The functionality of MMS

It is important first to explain the derivation and importance of MMS in more detail than the description in section 4.2.3. As was explained there, the application layer of MAP has a file transfer service, FTAM, and a messaging service. FTAM's main use will be to transfer files between minicomputers connected on the factory broadband backbone. The messaging service too has a file-transfer facility to transfer files between controllers, minicomputers and other shop floor devices.

No satisfactory standard message format existed when General Motors was carrying out its early MAP implementations. So GM, in collaboration with Allen-Bradley, Texas Instruments and a number of others, devised a variant of the standard message format and called it the manufacturing message format standard (MMFS or 'Memphis'). MMFS was implemented in MAP 2.1, still the most common implementation and the only one for which products are available until mid-1988. MMFS provides a standard syntax for messages between factory devices. MMFS assembles frames of data into complete messages which can be passed between numerical controllers, robot controllers, programmable controllers, automatic guided vehicles and computers of various types. An MMFS message consists of a series of fields, or complete data blocks which, taken together, form one of a number of MMFS PDUs. MMFS defines what the fields should contain and in what order the fields should be assembled in each PDU. In MMFS, certain parts of the MMFS syntax apply to certain devices – programmable controllers, robots and so on. Indeed, MMFS contained a very complete – if sometimes ambiguous – description of the messages and relevant commands that would be needed for all sorts of factory floor equipment. No particular application could possibly want more than a subset of these. So devices are loaded with the minimum subset each device needs to be able to function effectively.

Within its limitations it has worked very well. But GM and its collaborators knew it would have to be improved. Parts of it, for example, are open to a variety of interpretations. It also uses a non-preferred syntax (language). As explained in section 4.3 on the presentation layer, MMFS used a non-standard syntax instead of the ISO's Abstract Syntax Notation One (ASN.1). ASN.1 is a slight variant of the X.409 syntax recommended by CCITT. So GM and the others submitted MMFS to a committee of the Electrical Industries Association, committee 1393, for revision. The 1393 committee formed subcommittee 1393A to do the work. They re-wrote MMFS in the ASN.1 syntax and made other quite extensive changes. The result is variously known as EIA standard RS511 or as the manufacturing message service (MMS). Draft five of RS511 was sent out for voting in June 1986 and accepted in September that year. RS511 went into its sixth draft in the early summer of 1987.

MMS also assembles frames of data into complete messages. It also defines the message notation. This is important because the message notation contains information about the purpose, length and value of each element in a message. MMS describes the range of messages available both to the application processes and to other SASEs (see section 4.2.4). And it contains information about what should happen when a message is received. MMS operates by exchanging pairs of messages – one in each direction – between computer and factory item. In this way cell computers can exchange files from one to the other, and one of the cell computers could use these messages to change part of a robot's control program. Or a machine-tool controller could exchange data with either its cell controller or a PLC connected to a nearby conveyor. Or the computer can ask the controller to use an MMS subroutine to ask another device for information which, when the second dialogue is over, can be read back up to the computer which made the first request. The service allows a number of important factory floor functions such as remote control, access to user or part-program files and so on. This functionality is one of MMS's most important features, as will be explained shortly.

The dialogue begins with a request for information, perhaps about the state of a 'variable' – MMS defines 'variable' very widely, so it could be a file or other data – and the second message maybe either the value of the variable, which is read back to the initiator of the dialogue, or a confirmation that the variable has been read. MMS is a connectionless protocol: that is, it does not use any acknowledgement of the receipt of a message, so it assumes the data transfer is error-free.

MMS is probably the key component in the MAP specification. MAP without MMS is a mere communications system. But users and vendors who understand MMS and apply it properly will have an immensely-powerful – and genuine – computer-integrated manufacturing (CIM) tool at their disposal. Within MMS lies the potential for vendors to supply standard black boxes for control and other shop floor functions. These boxes will differ only in some machine-specific interfacing memory. They can be plugged together and a near-standard host program running on a factory-supervision computer will do the rest. Any piece of software which runs a standard operating system such as Unix will be able to interact through a logically-hierarchical data structure with any other process anywhere in an operation. MMS provides the messaging structure which makes all this possible and without which MAP is just a signalling system on a broadband cable.

This is because of the file transfer element in MMS. There are three ways MMS loads user programs into programmable devices on the shop floor. Two of these ways control the devices from a host computer. In these cases the host must hand the program to the device. So the host must have access to the file which contains the user program. But the host must also know other things about each device. For example, it wants to know how big the device's memory is so that it can decide accordingly whether to download the whole program or dripfeed it. These device characteristics will often be vendor specific, and having to program the host to allow for vendor specific device characteristics is hardly in the spirit of the MAP effort.

The third method is much more important to true CIM applications. In this, the host does not need to know anything about the programmable device on the shop floor. It does not even have to access the user program. It uses MMS to send a message to the device telling it to request a program and leaves the device to get on with it. The device could retrieve the program from either a local disc or, using MMS's built-in file-transfer protocol, from a remote database. The data transfer could take place whatever the particular type of local device, as long as it was a MAP-compatible device or had a MAP interface of some kind.

And, as in the previous cases, the load can be either the whole program – a 'static' load – or part of the program at a time – 'dynamic' loading. This means even simple machine-tool controllers can be used to execute large, complicated programs involving multiple cutters, probing cycles and so on by doing these programs a bit at a time. So relatively simple machine tools could operate unmanned, producing higher-quality parts. This is the thinking behind the RC4, a product which Renishaw of the UK launched at the 1987 Automan show in the UK.

But there is a danger that some of the attractions of MMS could be diluted, or lost altogether. MMS's file-transfer mechanism is fairly simple. It only allows the transfer of what are called unstructured binary files but, on the shop floor, these are the only files that need to be transferred. However, the other file transfer protocol, FTAM, can transfer the more complex files which are routinely exchanged between minicomputers. FTAM will be necessary to implement TOP but it is over-complicated for MAP. And it cannot be used with the high-speed, enhanced-performance or 'collapsed' architecture version of MAP which will be used for real-time control.

Nevertheless FTAM is a draft international standard, DIS8571. MMS has been adopted by working group five of the International Standards Organization's technical committee TC184 as draft proposal DP9506. But MMS has much further to go than FTAM before MMS becomes an international standard and so may be considerably revised. The second draft of DP9506 had yet to go to ISO by the spring of 1987 because it was still being drafted in the 1393A committee of the US Electrical Industry Association as EIA standard RS511. Most of the practical work on MMS, including CNMA's (see section 7.8), has been based on 1393A's fifth draft of RS511. RS511 went into 1393A's draft six in the spring of 1987 and 1393A chairman John Tomlinson of GM says it may go into a seventh or subsequent draft before it goes back to ISO for its second ballot as DP9506. At this stage the draft will be frozen for two months while member countries of ISO study it and vote accordingly. If the draft is voted down the vote may have to be repeated. After this, even if ISO accepts DP9506 and it becomes draft international standard DIS9506, it will still need to be refined in committee.

The point of all this is that the MAP 3.0 specification to be published in 1988 will include whatever version of RS511 finally comes out of 1393A. Depending on how the drafting process goes, RS511 could result in a MAP 3.0 specification which is stable for at least three years: or it could result, thanks to the voting on DP9506, in the arrival of a MAP 3.1 a matter of months after the publication of MAP 3.0.

6.1.1 CNMA's concerns

The most rigorous examination of the strengths and failings of MMS carried out so far has been done by the partners in ESPRIT project 955, the communications network for manufacturing applications (CNMA). For over a year the CNMA consortium has worked on project 955 of the European Community's ESPRIT programme. CNMA has put MMS (RS511 draft five) through all sorts of practical hoops – including the development of MMS conformance tests – so that it could be used in a MAP and TOP demonstration cell at the Hanover Fair in April, 1987. At that point CNMA revealed its interim conclusions about the way MMS should develop as it goes through the drafting process.

Two points in particular worry CNMA. One is that too much might be added to the basic MMS subset. 'MMS ... should not be extended unless absolutely necessary.' CNMA found the basic subset of services quite adequate to support many different applications even without the specific support for robots and other devices. So one unresolved issue is what use to make of the messaging service within the MMS sub set after the companion standards are finalized. The intrinsic MMS messaging format can be used for robot and NC control, some say very effectively, but its use may be frowned on when the companion standards eventually arrive. What happens to those users who have gone ahead with systems which use the intrinsic, not the companion, messaging formats? Will companion and intrinsic formats interwork?

The other worry is the opposite: that too much might be cut out of the MMS subset. CNMA is concerned that the MMS file-transfer service will be cut out in favour of FTAM. CNMA's interim report says: 'This would require another protocol machine to be implemented which would increase the cost and software complexity and would not be suitable for the collapsed architecture environment.' Specifically, CNMA wants MMS to retain its file transfer function. This is a much more serious matter because the loss of MMS file transfer would undermine MMS's usefulness in CIM, for all the above mentioned reasons.

MMS, at least as drafted by the early summer of 1987, had weaknesses as well as strengths but, before outlining these, it is necessary to explain a key difference between MMS and its predecessor MMFS which has a bearing on possible migration paths between the two.

6.1.2 Migration

The watchword of the MAP/TOP community is that there will only be one version of MAP and one version of TOP, worldwide. Inevitably, in the still-continuing development phase of the two specifications, different versions have emerged. In the early 1980s, GM's first priority was to get a broadband system up and working regardless of what OSI standards were available. As the standards have gradually come out, the earlier versions of GM's MAP implementations have become out-dated. Similarly, when demonstrations were needed of MAP, at the National Computer Conference in Las Vegas in 1984 (see Chapter 1) and at Autofact in Detroit in 1985, the most important thing was that the demonstrations should be seen to work, not that they should be OSI-conformant.

These circumstances led to the development of the so-called 2.x versions of MAP, and there are even variants from it such as Autofact MAP: version 2.1A is a corrected version of MAP 2.1. MAP 2.2 is a further amendment of 2.1A which added the EPA, carrier band, more detail on bridges and their use, and a new appendix on naming and addressing. It also added some information about how to make MAP 2.0 systems interoperate with MAP 2.1 and 2.2 systems: a main difference concerns the use of the connectionless network protocol (ISO 8743). This is inactive in 2.0, being replaced by a string of 0s in the network-layer header. But 2.1 and 2.2 systems have a full header as well as this inactive subset and this means they can communicate across more than one network.

In the beginning, the intention was to make all the 2.x versions upwards-compatible with each other. This means that users of 2.0 would be able to upgrade existing equipment to 2.1, and 2.2 would be designed in a way that would enable users of 2.1 to upgrade to 2.2 and so on. The original thinking also ran, however, that version 3.0 would be definitive, as perfect as it could be made, but not necessarily reached by an upgrade from 2.x. As late as August 1986 GM's staff were referring in correspondence with network users and manufacturers to 'a period of upward compatibility', implying that upward compatibility would only last a finite time. In other words, 3.0 would not be compromised by the need to allow users to reach it using their older systems.

This original philosophy now appears to have given way to the idea that hardware, software or both will be made available which will allow vendors and users to convert from 2.x to 3.0. Towards the end of 1986 the US MAP Users' Group set up a working group to find ways of providing upgrades from 2.x to 3.0 and a paper was due for publication during 1987 outlining how this would be possible.

6.1.3 Companion standards
One feature of MMS makes it hard to see how a migration path can be easily developed. MMFS was ambiguous in some ways but it did specify all the commands available – even to device-specific commands for robots, programmable controllers and so on – and the expected response to those commands. But MMS takes the form of a skeleton standard which makes no reference to specific devices such as robots or programmable controllers. It is a generic standard for any device on the factory floor.

The relationship between the generic MMS and the specific requirements of particular pieces of equipment will be established by other US industry associations, which will provide specific 'companion' standards of their own. Thus, for example, the Robot Industries Association (RIA) of the USA has been asked to draw up a companion standard to MMS. So drawing up this companion standard is now the work of RIA committee R15.04. (The US bodies have committees to themselves at international level. The RIA forms sub-committee SC2, of ISO TC184 and the EIA numerical-control committee corresponds to SC1 of TC184. The Instrument Society of America, which looks after the interests of the process-control industry in the USA, is also involved. But the ISA's participation relates to standards drawn up or about to be drawn up by the International Electrotechnical Commission.) The addition of these companion standards will not be too disruptive, partly because they will use the same syntax notation (ASN.1) as MMS itself. But it should by now be clear that the structures of MMS and MMFS are quite different, and there is some scepticism about how easy it will be to provide upgrade paths from one to the other. Some in the MAP task force say translation can be demonstrated for a dozen messages but a fully interworking set of commands is a different matter.

The point of going into all this is to outline the immensity of the problem faced by anyone drawing up a design for a system which will, without too much difficulty, convert MMFS-based systems to MMS.

6.1.4 Ambiguities
From the strengths of MMS to what some have seen as a weakness. As

explained above, committee 1393A has left much of the interpretation of MMS to other standards-making bodies. Some MAP students, however, have complained that the way 1393A drafted the MMS 'core' makes MMS very easy to use but may lead to different interpretations of it. MMFS did have the advantage that it was very specific about how the services contained within it were used.

For example, CNMA workers have found in using MMS draft five that, although MMS provides for a number of services – 'file obtain', 'file read', 'file transfer' and so on – it does not explain how to use them. For example, in operating a conventional numerically-controlled machine tool, the operator will press 30 or 40 buttons on its control panel. The series of steps the machine must take to make a part is contained in a part program which is normally in the form of a paper tape or its equivalent. Some of the buttons are used to load the program into the machine's memory. Once a program has been loaded into a machine, other buttons have to be pressed to tell the machine what to do with the data in the program. The operator uses the buttons to turn the machine on and off and to tell the machine which tooling to use at particular parts of the program, when to load and unload parts and so on. Any fully-automated system which seeks to operate the machine remotely must send the machine a program able to emulate the operation of these buttons. MMS is capable of loading the equivalent of the tape into the machine. But, in its present form, MMS is unable to press the 'start' button or tell the machine what kind of tape it has stored: a tooling tape or a part-program tape and so on. MMS specifies none of this.

There is a way round the problem. Applications software which supplies the missing information can be written for either end of the link between the host computer and the machine tool controller. But MMS sets out no way of approaching this task, so there is a danger – which has been described as 'a confident prediction' – that every controller manufacturer will do it a different way from his competitors. The content of part programs already differs from one manufacturer to another but at least the files can be moved about. The provision of different applications software for direct numerical control links would make it impossible even to transfer files. If the ambiguities in MMS are allowed to develop then MAP, far from offering a multi-vendor solution to factory communications problems, will be a step backwards.

These already-obvious problems are only the beginning. Apart from the inclusion of MMS, MAP 3.0 will, as noted, have a presentation layer, hitherto absent in 2.x. The final version of TOP 3.0 will be published at the same time as MAP 3.0. But 3.0 is not the end of the development of MAP and TOP. Future versions of the specifications are likely to contain provision for fibre optic connections, over-the-air networks and connectionless upper-layer protocols. The use of a connectionless network protocol in the Internet sublayer has already been mentioned (Sections 4.6 and 6.1.2). There is also a proposal to provide a connection-oriented version of this protocol at some time in the future. As already noted, there is much work still to do.

6.2 Interoperability

There is one more small point to mention on the compatibility front. It was only at the end of 1986 that manufacturers of MAP interfaces were able to say with any confidence that their systems would be able to work with those from other manufacturers. When MAP was in its early stages, the equipment used was supplied by Concord Data Systems. CDS's equipment ran at about 5Mbit/s. It lost ground when the proposed speed of MAP broadband went up to 10Mbit/s. This allowed Industrial Networking Inc, a joint venture between General Electric of the USA and a US communications company called Ungermann Bass, to steal a lead. CDS, which has now split off its MAP systems subsidiary as Concord Communications (CCI), has since made up lost ground. Some manufacturers of factory computer equipment, DEC, for example, use CCI interfaces and head ends and others, such as IBM, use INI. But, although both makers' interfaces are said to conform to MAP, CCI interfaces have had to use CCI head ends and INI interfaces have had to use INI head ends. The difference concerns the frequencies used in each makers' device. It means that the connection of a device with an INI interface and one with a CDS interface means that the user needs two head ends, one for each type of interface, and a router to connect the two (effectively separate) networks together. DEC and IBM systems were linked at the UK CIMAP demo through an Intel router. General Motors at Saginaw, for example, has used CCI token interface modules for the whole experimental plant to get round this interoperability problem.

At the end of 1986 CCI and INI were able to connect systems systems together which could interoperate, though this is different from having commercially-available systems which do the same. This potential problem explains what interoperability testing is all about.

6.3 Testing

If MAP is to become a single, international specification for factory communications the equipment said to conform to it must be rigorously inspected to make sure that so-called 'MAP' products really are MAP products. Exactly the same is true of TOP. This means that test centres must be set up around the world. Setting up test centres is difficult enough: who, for example, is to pay for setting up the centres, even if selling certified test software, say to vendors, becomes a viable proposition? But there is much more to testing than meets the eye.

Each test centre must test products for conformance in such a way that each test centre does its tests in exactly the same way as every other. This means the tests each centre uses must be exactly the same. This implies using software-driven tests from a single set of test programs instead of having testers at either end of a phone shouting 'hit return!' at one another. The test software runs the tests in a particular sequence and has built-in automatic evaluation. Tests are done one layer at a time, though multi-layer tests have also been proposed. The test software operates from the layer above the one tested and throws errors at the tested layer.

Providing a single set of tests also implies the use of 'portable' software, which means that the software must operate on the widest possible number of systems. In effect this means using software which runs on a standard Unix V or Berkeley 4.2 operating system. The test set 'authorized' so far is that from the Industrial Technology Institute in Ann Arbor, Michigan, USA. ITI's software was developed for the Autofact 1985 demonstration on a DEC Vax 11/750 using as an operating system a Unix 3.6 emulation itself running on a VMS operating system. This was not satisfactory. ITI has now ported its software to run on genuine Unix Berkeley 4.2 on a Sun workstation.

ITI is not the only organization which is developing test tools – the Fraunhofer Institut of Information and Data Processing is also doing work under the communications network for manufacturing applications project 955 of ESPRIT (see sections 6.11 and 7.8). But, for the moment, the ITI is the only one recognized by the US MAP/TOP users' group, even though it provides only MAP, not TOP, test tools.

Having tests which are exactly the same everywhere also implies that the results of the tests should not be open to different interpretations by either different test centres or even different individuals within the same test centre. This is a matter of continually refining the specification as well as the set of test tools to eliminate ambiguities. So far these have cropped up deep in the technicalities of MAP and TOP. Engineers talk about variable 'time-outs' and 'slot-times'. The values of various parameters can be adjusted but some values genuinely are options and others, though they can be adjusted, are not really optional because the specification defines exactly which options are used.

Conformance testing is only part of the testing problem. For even if products conform to the MAP specification, they may not interoperate. An example of this kind of problem was mentioned in section 6.2, where INI interfaces needed to use an INI head end and Concord interfaces needed to use Concord head ends. Both INI and Concord interfaces and head ends conformed to the same then current version of the MAP specification but they would not interoperate with one another.

Interoperability tests have to confront the problems of interoperation at all seven layers, not just the physical-layer problems that confronted Concord and INI. What is more, to test that two systems work together, though complicated, is fairly simple because there is only one pair of interacting devices. If three systems are available, there are three sets of interactions between pairs of devices. If four devices are available which may be used on the same network, the number of possible interactions goes up to six. Ten available devices from different makers puts the number of possible interactions up to 45 and so on. This is why MAP users so far have avoided any of these problems by choosing one interface supplier and using that interface for everything.

Two good examples here are General Motors' own highly-automated Saginaw 'Vanguard' pilot and its GMT400 truck and bus group plants. Saginaw uses Concord interfaces, GMT400 uses INI. Moreover, it is impossible to identify all the possible circumstances which might arise in interactions among these disparate systems, so it is also impossible to predict how these interacting systems would behave. All that can be done is to identify specific core interactions and see what happens in those circumstances. As experience of the use of these systems grows, core interactions can be modified or added to the tests.

Testing for interoperability is not just a matter of linking two or more systems together and seeing if they work. That is just functional testing, the kind of thing that, so far, is all that has been possible for demonstrations like Autofact and CIMAP. All the tests have error-recovery at their core. True testing means hitting the devices under test with specific errors and seeing how they recover from those errors. These error-recovery tests must be controlled and they must be repeatable. In the case of interoperability, this means connecting two devices together, hitting both with errors and observing the results. Even interoperability is not the end of the matter. MAP and TOP systems are bound to develop as the technology develops. One obvious change this will cause is an increase in the speed of the systems. Tests need to be developed for system performance as well as conformance and interoperability. So far, most progress has been made on conformance testing. Hardly any has been made on interoperability except in specific cases for specific purposes, such as demonstrations at CIMAP and Autofact, for example. And no-one has given much thought to performance testing except to agree that it would be a good thing.

So for the moment, the only tests we can talk about are those for conformance and for specific cases of interoperability. Even the important conformance tests were not ready by the middle of 1987. How could they be? The conformance tests that count, in MAP's case, are those for conformance to MAP version 3.0, not its 2.x predecessors, though the development of 2.x test tools will have allowed some of the MAP and TOP support community to become familiar with what is involved. The preliminary draft of MAP 3.0 was not published until April of 1987, so tests for conformance to even this preliminary draft could not be available until well after that. And the final revision, the first 'final' version of the specifications, may look quite different from the preliminary version. The final version of MAP 3.0 will not be published until at least April 1988. If the specification changes between preliminary and final MAP and TOP 3.0, as it is bound to, the tests will have to change too.

So although some companies have committed themselves to delivering their first MAP 3.0 products a certain time after the specification is published, users should make sure that, when this happens, the tools to test these products have also been available, and that the products have been tested in accordance with them.

A key organization in all this is the US Corporation for Open Systems (COS), which oversees MAP testing in the USA. COS, a non-profit making collaboration among 61 member companies, is concerned not just with MAP and TOP but with open systems generally. Its brief is to accelerate the international introduction of inter-operable, multi-vendor products and services operation under agreed OSI, integrated services digital network (ISDN) and related international standards. (ISDN is an effort to eliminate the use of separate networks for voice, telephone, facsimile transmission, telex and all other communications. It is a telecommunications concept and, though it may affect or interact with the MAP/TOP initiative, as such a concept, it is beyond the scope of this book). COS has identified certain key parts of the seven layer model which are common to those systems which adhere to the OSI model. It has agreed to carry out or subcontract testing for these key parts, which are collectively known as the COS 'platform'. The other tests, those specific to MAP, TOP or other specifications, have to be devised as their promotors see fit. The US MAP/TOP Users' Group (UMUG) has taken on the non-platform tests for MAP and TOP.

Some friction has developed between UMUG and COS because UMUG feels that the COS platform is so restricted; UMUG also wants to make COS move faster. By the beginning of 1987 about 20 parts of the MAP and TOP specifications had been identified which need testing but which COS had excluded from its platform. The COS platform includes layers five to seven, the file transfer access and management (FTAM) service, and the message handling service. But this excludes some of the most difficult areas of the MAP seven layer stack. It excludes the whole of layer four – the transport layer, which has a lot of built-in ambiguities – and the complex network layer at level three.

It also omits the MMS, which is due to replace MMFS in the top layer of the seven layer model. Another omission from the COS Platform is the enhanced performance architecture. COS will not test the logical link control sub-layer or any other sub-layer in layer two. The US Institute of Electrical and Electronics Engineers is working on tests for the trickiest layer, layer one.

There were problems even with the tests which were in the platform. COS sent out requests to any subcontractors interested in developing tests for layers three to seven of the seven layer model. One request was for tests for layers three and four, and two more requests went out for layers five to seven, using FTAM and the message handling service. COS was expected to announce its sub contracts in January 1987 but the announcement did not appear until three months later. One reason is that COS did not get the expected response, especially to its requests for layer three (network) tests.

The test programme bears heavily on the plans for the US MAP 3.0 demonstration, the Enterprise Network Event, in Baltimore, in June 1988. The first draft descriptions of the conformance test tools will not now be published before the late autumn of 1987 and will have to do the rounds before they can be considered complete. This will take until the beginning of 1988. Then interoperability tests will be developed at seven pre-staging sites, one of them at the Production Engineering Research Association (PERA) in the UK.

Alternatives are being considered to get round the need to develop completely tested products by the ENE deadline. One proposal is to use green and gold dots to show which products are safe to use (gold dots) and which have some further testing to undergo (green). This may run aground because of the legal risks involved, particularly on product liability. In US law, some risk may attach to those who tested the products as well as those who used them.

Since, in general, the availability of conformance and other tests determines the rate at which MAP and TOP products come onto the market, it is now inevitable that the first MAP products conforming to the third version of the MAP specification will now not be available until at least the middle of 1988. The development of test for TOP systems has been slower than those for MAP. This means TOP has had to be tested using MAP test tools with different software at the lower levels.

As already mentioned, so far ITI is the only authorized test house. However, the introduction of others is vital to the MAP and TOP programme provided that their operation is managed and controlled. The development of test tools and other matters will now be co-ordinated by the MAP/TOP user groups and by the US Corporation for Open Systems. This will be one of the tasks of the COS-MAP/TOP joint advisory council, which first met in February 1987.

7
MAP AND TOP APPLICATION CASE STUDIES

Introduction

Although MAP has to go through a few more hoops before it can be regarded as a mature technology, many UK and US companies are trying hard to make themselves thoroughly familiar with it. As pointed out elsewhere in this book, many have taken part in exhibitions and demonstrations in Europe or the USA and others have installed MAP and TOP systems in their own factories.

In this chapter the activities of some of these companies are described. Some of the descriptions, for example General Motors, ICL and British Aerospace, have been compiled from visits to the plants and/or conversations with those implementing the systems. In other cases the material has been supplied by vendors. All the material either illustrates a particular approach or points the way to the future.

7.1 The Towers of Hanoi and Beyond

This case study details the approach taken by Tandem that enabled them to demonstrate MAP at Autofact 85, the lessons learnt from that exercise and the development that has taken place since.

7.1.1 Autofact 85

The Autofact 85 exhibition was held at Cobo Hall, Detroit, USA, during November 1985. Organized by the Society of Manufacturing Engineering, it was primarily a computer aided design and manufacturing show. A section of the exhibition was given over to MAP and TOP where a number of demonstrations could be seen.

MAP is a method of communication and since it has no visual elements it is usually demonstrated by showing how a task could be achieved while using a MAP local area network (LAN). Tandem's MAP demonstration was no exception to this. It showed how the 'Towers of Hanoi' puzzle could be resolved using a vision system to look at the puzzle and a robot to play it. Towers of Hanoi is a game in which the object is to move a set of discs to one of three posts, where each disc is a different size. There are two rules, move only one disc at a time and a larger disc cannot be placed on a smaller one.

The demonstration would begin by inviting an exhibition visitor to participate in the Towers of Hanoi game. The visitor would be asked to set up the discs and then select which post the discs should be moved to. After set up, the vision system scanned the game and passed details of each disc's position over a MAP LAN to a computer.

Based on the information from the vision system, an application program running on the computer determined the moves that would get the discs onto the required post. Having solved the puzzle, the computer passed the necessary moves back through the MAP LAN to the robot. The robot then moved the discs, one at a time, to complete the game.

In essence, the demonstration showed a shop floor device (the vision system) communicating via a MAP LAN with an application program and that application program communicating over the same MAP LAN to another shop floor device (the robot). For the team responsible for building the demonstration system the ability to install MAP software over a number of communicating devices proved to be a key factor in the MAP implementation.

The demonstration team, led by Andy Poupart with much co-operation from other suppliers, designed and implemented the system (see Fig 7.1) in nine months. The computer, a fault tolerant system, was connected to a Packet Multiplexor (PMX) over a 56kbit/s synchronous link using a proprietary protocol. Fault tolerance was achieved using a standard Tandem EXT two processor system. The PMX is a specialist communications processor that allows data to be exchanged between the computer's communication protocol and a modified high level data link control (HDLC) protocol. On the HDLC side the PMX communicated with a Concord Data Token Interface Module (TIM) at 56kbit/s. A TIM is a device that enables equipment to be attached to a token bus network, in this case a MAP LAN. The TIM thus converted data between the modified HDLC of the PMX to the IEEE 802.4 protocol of the MAP LAN. The MAP LAN itself was a coax cable (an inner, information carrying cable, screened from outside

Technical details

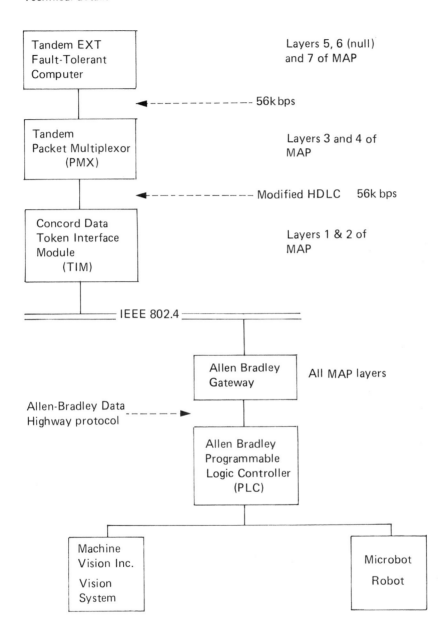

Figure 7.1 The Tandem MAP demonstration system

electromagnetic interference by an earthed sheath) with a 300Mbp/s bandwidth of which MAP used 10Mbp/s. Connected the MAP LAN was an Allen-Bradley gateway which communicated to a programmable logic controller (PLC) using the proprietary Allen-Bradley protocol 'Data Highway'. The PLC then communicated with Microbot's robot and Machine Vision's vision system.

The MAP software was distributed across this hardware with various devices handling different layers of MAP. The computer ran programs for the application and session layers (the presentation layer was NULL, which means it is unfilled) while the PMX handled the transport and inactive network layers and the TIM looked after the physical and datalink layers. On the shop floor device side of the MAP LAN, the Allen-Bradley Gateway was responsible for all MAP communications.

Machine Vision supplied and programmed their vision systems to scan the Towers of Hanoi game and pass information about which discs were on which posts. The vision system is more normally used for tasks like detecting flaws in paintwork. Microbot supplied and programmed their robot with a set of programs to move a disc from one post to another and Allen-Bradley also assisted by programming their PLC to handle the robot and vision system.

As a result of the experience gained in building the demonstration, Tandem was able to submit its transport layer for conformance testing and became the first vendor to obtain the International Technology Institute certification for MAP 2.1 at the transport layer.

7.1.2 Lessons learnt
The system on the day worked impeccably. It is our belief however, that there are three major problems with the approach taken to the demonstration:

Bandwidth
The bandwidth of the system is dependent on the bandwidth of the slowest process. The demonstration had a computer communicating to a PMX and the PMX communicating with a TIM. Both of these communication links were running at 56Kbp/s while the MAP LAN runs at 10Mbp/s. There is clearly a bottleneck using this approach since the MAP LAN is capable of operating almost two hundred times faster than the PMX can pass information to (or from) it.

Reliability
The approach used in the demonstration resulted in a system that was vulnerable to single points of failure. Should a failure have occurred in

say the PMX to TIM link, the ability to communicate over the MAP LAN would have ceased. This is embarrassing at a demonstration but potentially disastrous in the real world.

Flexibility
The use of other vendor's hardware tied Tandem into a particular development route that they had little control over. Furthermore with the distribution of the software over a number of devices not directly under the computer's control, system/network management became a problem, for example it was difficult to obtain retransmit statistics.

7.1.3 The way ahead
One site that will be testing the results of MAP research is the CIM laboratory of the Arizona State University (ASU). Directed by Dr Daniel L Shunk, the laboratory's goal is to help manufacturers increase productivity and maintain a competitive edge through efficient automation. A Tandem computer with Factorial's Paperless Factory software has already been installed and is contributing to the work of the laboratory. The University has been working on MAP research for some time now and has installed an extensive broadband cable system (see Fig 7.2) and has plans to integrate a number of different computer systems into its network.

Finally, it should be remembered that MAP is an enabling technology; it allows devices and applications on various machines to communicate with each other. Note the term 'allows'. For an application to usefully communicate with a device or another application it presupposes that the application is capable of sending data to the MAP processes. Also that the data being sent is of value to the target application or device. For example, an NC lathe will be more than a trifle confused on receiving purchase order actions from a material requirements planning (MRP) system. To use MAP 'in anger' requires not only the investment in MAP hardware and software, it requires investment in MAP compliant applications and in the intellectual exercise of what talks to what – and why.

7.2 Cell controller applications

Introduction
This case study describes the running of two Gould Cell Controller applications. To date, Gould has shipped two FM1800 Cell Controllers to customer sites where installation has been completed and the systems are running the applications for which they were purchased. In addition

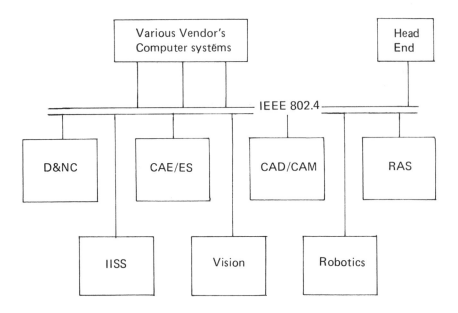

Figure 7.2 The planned ASU network. It will support the work of a number of laboratories: database and network (D&NC), computer-aided engineering and expert systems (CAE/ES), computer-aided design and manufacturing (CAD/CAM), reliability, availability, serviceability (RAS), integrated information support systems (IISS), computer vision and robots

there have been a number of other shipments to beta sites, test installations and to customers where installation is either incomplete or not yet started.

The Gould Cell Controller (GCC) is aimed primarily at four major application areas:

1. Flexible manufacturing cell control
2. Data collection, concentration and manipulation
3. Batch process control and monitoring
4. MAP gateway to multiple dissimilar programmable devices

The following summarizes the two installed systems, giving a short description of the application, configuration of the system, problems encountered during design and start-up and the benefits gained. The two systems in question cover the first two application areas above.

7.2.1 Application 1

Application description
This application involved the assembly of transmission system components for a major automotive manufacturer. The facility was part of an experimental/pre-production line itself based on an FMS concept. The Gould Cell Controller (GCC) was responsible for scheduling the activities of the cell, the co-ordination of two multi-axis robots and a part positioning local and remote operator interface, and reporting cell status, progress and diagnostic data to the area minicomputer system.

System configuration
The system configuration is shown in Fig 7.3 and consists of one GCC, two GM-FANUC robots, one Gould 984 Programmable Controller (which controls the part positioning unit), one Gould IM1062 operator interface/ graphics display system and one IDT (Industrial Data Terminal) operator interface unit.

Communications between the GCC and each robot is via a MAP 2.1 10 Mbit/s broadband cable link – both the GCC and GMF Karel controllers providing a fully integrated MAP interface. Communications to the 984 PC and the operator workstations is via Gould's Modbus LAN at 19.2 kbit/s . Communications to the area computer is also via MAP.

Problems encountered
The system is generally working well, in line with expectations. The usual commissioning difficulties encountered with a new system of this nature were overcome much as expected. Whilst this is the case it should be noted that the time taken to translate messages from MMFS to other protocols would be a problem if very high data throughput was required. Whilst MAP provides a clean multi-vendor interface in currently available products this seems always to be accompanied with performance limitations where speed is concerned. The use of the MAP enhanced performance architecture or Mini-MAP based products can overcome this and it is recognized that this is the preferable networking option at levels below the cell level in the factory structure.

Perceived benefits
The use of MAP communications between devices of dissimilar types and protocols gives the direct benefit of a clean interface with minimal start-up problems. The use of the Gould Cell Controller gives a further benefit in that the cell co-ordination is provided by a system specifically designed for that type of operation particularly from the software viewpoint. The IEC standard Sequential Function Chart/Function Plan

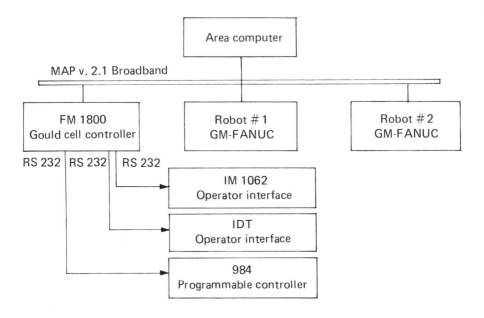

Figure 7.3 System configuration of the Gould Cell Controller for flexible manufacturing cell control

language combined with the option of 'C' language subroutines provides a unique mix of control oriented and high level languages ideally suited to this type of application. This kind of approach is more readily understood by engineers and technicians used to factory floor devices and reduces the dependence on programmers unfamiliar with control systems which the use of general purpose computers often brings.

7.2.2 Application 2
Application description
This application reflects the second major application field listed for the GCC ie data collection, concentration and manipulation. The GCC is configured as a data concentrator between Gould Programmable Controllers and an operator interface/graphics display system. The process is a spent foam recovery system in the casting line of a large foundry operation. Foam which is used in the preparation of sand moulds is displaced during the casting process and recovered by a number of PC controlled operations. The GCC provides a peer-to-peer

MAP and TOP Application Case Studies 123

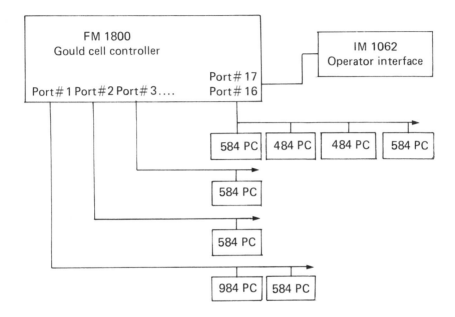

Figure 7.4 System configuration of the Gould Cell Controller for data collection, concentration and manipulation

service for information transfer between the PCs as well as collecting process alarm and other diagnostic information.

System configuration
Figure 7.4 shows the system configuration. The Gould Modbus protocol is used across an RS232 link in each.

Problems encountered
Although MAP communications are not actually used in this application it should be noted that the GCC still translates from the specific device protocol into MMFS which it uses as a device independent standard protocol. As with the first application the ability to hook up devices is superb but the penalty at present is the performance limitation imposed by the protocol translation which is still carried out to the MAP specification.

Perceived benefits
The GCC has been uniquely designed to provide for real-time serial I/O

capability. As such, it can provide a real-time front end to a traditional minicomputer or graphics display system. It can collect information, condense and operate upon it, and, if required, transfer the results at MAP network speeds (10Mbits) to other host systems. This allows these hosts to perform higher level functions more easily. Another example would be the connection of 60 Gould Programmable Controllers operating as Modbus slaves to the GCC. The GCC could connect as few as 5 PCs per Modbus network, and still have four ports available on the GCC for connecting printers, CRTs and other devices. Local display of information would be possible, as well as transfer of condensed information upwards across the MAP networks. Program upload or download of the PCs is also available in this configuration.

7.3 A link from design to assembly and inspection

The following case study describes the design and automated manufacture of a real-world component – a car water-pump – in conjunction with the latest techniques in machine vision. This demonstration cell (Cell 7) at the CIMAP Event, was the result of a partnership between Reflex, Motorola and ICL. A fully integrated design and assembly system was shown in operation (shown diagramatically in Figure 7.5). It illustrated how CADCAM systems can communicate with, control and monitor an intelligent assembly cell, via TOP and MAP networks.

There were five standard pump designs, each pump body being made up with one shaft, seal and impeller. A number of component parts are interchangeable between pumps so that over a dozen possible combinations could be achieved to demonstrate flexible automation.

Design was performed, and pump configuration defined, on an ICL graphics workstation, running DIAD 2D/2.5D drafting package, which communicated via the TOP network to a second ICl graphics workstation running GNC software. This package is for preparing numerical control data. This workstation sent geometric data, giving the mechanical details of the assembly plus assembly instructions, over the TOP network, via a router, onto the MAP network and to the Motorola System 1131 – a 32-bit UNIX box based on the MC68020 processor. This box had a standard system V/68 UNIX implementation, and contained the MVME372 MAP card and the MVME371FS broadband modem – everything required to interface to a MAP network.

All seven layers of the MAP interface for the Reflex demonstration are implemented by the Motorola's MicroMAP software, which is fully

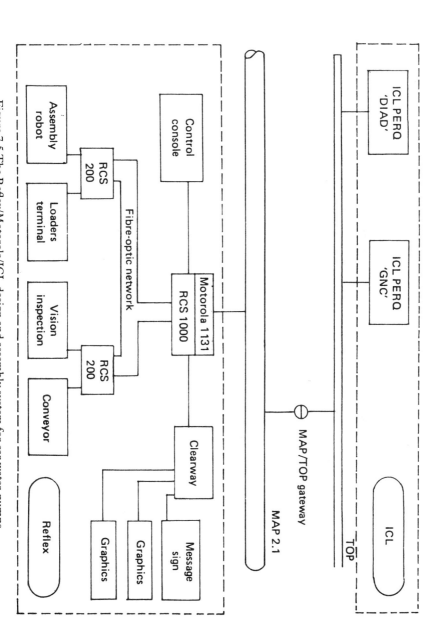

Figure 7.5 The Reflex/Motorola/ICL design and assembly system for car water-pumps

conformant to the latest MAP 2.1 specification. MicroMAP runs on Motorola's NVME372 board, which has its own MC68020 processor plus an MC68824 token bus controller chip. In addition, the Motorola system will carry out file transfer over FTAM.

From its interface, data is passed to the Reflex RCS1000 Control System. This advanced real-time intelligent controller features modular hardware and portable software to cater for a wide range of control and monitoring requirements. RCS hardware consists of a family of software-compatible microcomputer systems based on the Motorola M68000 microprocessor family. Each module can be used as a stand-alone controller or, for more complex applications, a number of modules can be networked together.

Software for the RCS family has been written in a high-level language and follows a systems modelling approach. For a specific system, a number of functional packages can be assembled into a core package, which is then tailored to suit the application.

The RCS1000 passes control signals to the Reflex assembly cell, which consists of a high-speed Reflex assembly robot, an Argus vision system and a Reflex intelligent conveyor. Data is also sent back from the cell, through the MAP/TOP network, enabling the ICL CADCAM workstations to keep track of assembly data.

Typically, data received by the RCS1000 will be issued as a set of commands. For example:

(a) Assemble water pump No 3 (pump body)
(b) Use shaft type No 3
(c) Use impeller type No 3
(d) Use seal type No 3
(e) Here is the robot program to assemble the water pump

The conveyor used to transport pallets around the cell was made up from a modular system which offered bypass, single gate, double gate, single line and return line building blocks. It transports the various parts of the example pumps (for example No 3) – bodies, seals, shafts and impellers – around the cell between the video inspection system and the assembly robot.

The Argus vision station inspects the constituent parts of the pump for presence and location on the pallet, and sends a signal back to the RCS1000 indicating correct/incorrect piece part. If the vision check is positive, parts are sent to the assembly station and the necessary programs downloaded to the robot. If any of the components are unsuitable for the specified assembly or if they fail the visual check, they will be re-routed to the manual station for attention.

The Reflex is a three axis pneumatically-driven cartesian robot. It has a maximum speed of 2.5m/s and features a repeatable accuracy of plus or minus 0.01mm. It has no rotating parts, being based entirely on linear actuators. All robot programs are downloaded from the CADCAM system.

The conveyor sends a pallet with shafts on it to the vision station. The vision station again checks that the pallet carries shafts, and identifies the position of No 3 shaft. This knowledge is also passed to the robot via its DNC interface from the RCS1000 controller. The conveyor transports the pallet to the unload station in the robot. The robot assembles the shaft into the pump using the downloaded program, modified with the positional-offset of the shaft determined by the vision system. A similar sequence is followed by the impeller and seal until the waterpump assembly is finished. The conveyor then removes the assembled pump from the build station, and transports it to the unload station.

The process is now ready to start over again. The application can be set on a sequence of builds that can be interrupted by manual input from the CADCAM system. Thus work is always being carried out, and the principles of flexible assembly can be proved.

This cell relates MAP and TOP to a variety of other communications systems. The robot and the vision system are linked to the assembly system controller over a Reflex fibreoptic network. The controller also has an RS232 interface to run office functions. These include cell-status displays and the transfer of graphics and text files within and between MAP and TOP networks.

These files are available for display, manipulation and printing. The RS231 subnetwork also contains a modem to link the cell over the telephone for remote diagnostics and control.

7.4 MAP in the electronics test environment

Introduction
Based on the problems of P-E Consulting Services' clients in the electronics industry this study identifies how OSI standards, and in particular MAP and TOP, could be used to improve system interworking. The study is not an actual implementation of MAP and TOP standards, but is a strategy indicating how they could be applied to resolve realistic interworking problems; as more MAP/TOP products become available the strategy could be implemented. The focus of the study is the automatic test and repair requirements of an electronics company.

The study

Modern high-speed automatic test equipment (ATE) tests large quantities of electronic components in a short time, producing vast quantities of test data. The data itself can be used raw as a basis of assessing the particular component. More importantly the combined data for a batch of components can be used to assess any variance or trends in the actual manufacturing process. If handled correctly this statistical data can be used as feedback to the manufacturing process, keeping it within the optimum boundaries for production. Moreover the statistical data for a large population may provide feedback on design parameters of the components. Overall it can be seen that the test process forms an integral part in the closed loop process of design, manufacture and test. This becomes more apparent as the production process moves up to board level, where components are assembled into printed circuit boards (PCBs). At this level of integration, correct tracking and analysis of component test results, and modelling techniques provide for another feedback loop to ensure that the final board performs within its specified parameters.

Figure 7.6 shows the data flow between the system involved in the manufacture test of the PCBs. This is the first level of data flow in the test environment. The manufacturing host receives the test results of the individual batches of components, and of the final board. Board failure may be the result of incorrect board assembly. It may also be the result of component problems. The components may each pass their individual tests. However the cumulative effect of small perturbations away from the specified parameters may cause board level failures. This has the implication on the manufacturing process and on the design of the PCBs. The feedback of the test results is therefore very important in tuning the manufacturing processes and on the potential redesign of the PCBs.

The diagnosis and repair of faulty boards returned to the manufacturer is a parallel requirement to the original manufacture and assembly of the PCBs. In its simplest form the repair of these boards is a stand-alone requirement. The board is tested in the normal way against the specified parameters. However, from the raw test data it is necessary to diagnose the faulty components(s) or assembly fault. The diagnostic software could be an integral part of the standard test software, or it could be separate, even on a separate system. In this case the diagnosis is produced by the test system and the results transferred to a workstation where they are stored in a repair database. Diagnostic information is then available to the repair engineer working at the workstation.

MAP and TOP Application Case Studies   129

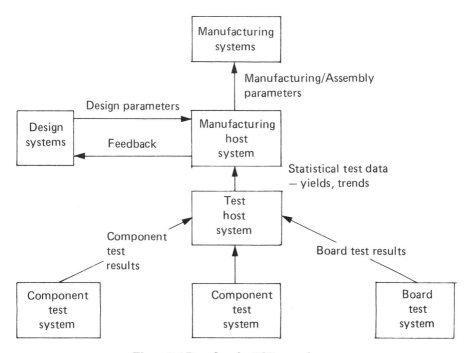

Figure 7.6 Data flow for PCB manufacture

However, such a system is not usually stand-alone. The test results are sent to the test system host for archival and further statistical analysis. Indeed the further analysis is extremely important in the feedback loop, to identify longer term problems in terms of mean time between failure of the PCB, and its individual components. The repair workstation would transfer repair data on the PCB to a repair host, which would have a master repair database. This database would hold the repair histories of the returned PCBs. Figure 7.7 shows the data flows between the system in the repair environment. However, the complete network has a far greater degree of interconnectivity.

There are a number of parallel networks consisting of the component and board test systems, the manufacturing host and manufacturing system, design system and the repair workstations. The intercommunication between these parallel systems takes place at a number of points. the component and board test system may be from different vendors. Each group may be centred on a test system host responsible for the collection and archival of raw test data. The raw test data could be reduced into statistical data by the test system hosts and sent to the manufacturing host. The manufacturing host is responsible for

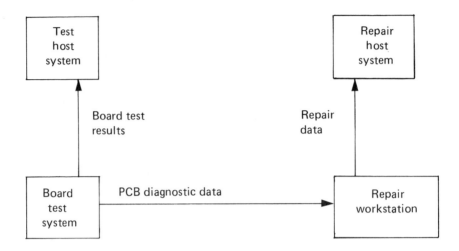

Figure 7.7 Data flow for PCB repair

maintaining statistical data for the test results from all vendors test equipment. It would also be responsible for work-in-progress and tracking of the manufacturing and test operations. Based on parameters received from the design systems it would also control the manufacturing processes through subordinate manufacturing systems. The manufacturing host system is the cornerstone of the whole operation and therefore is the key system in the data network.

The manufacturing host would also collect statistics from the repair host. By comparison of the test data for the faulty board with the original test data for that board, further feedback can be achieved in terms of both manufacture and design. The benefits of such a feedback system result in a more efficient manufacturing environment in terms of cost and quality, the two major requirements for a competitive manufacturer. The feedback can only be achieved with a truly integrated data communications infrastructure allowing the connectivity of all systems.

Figure 7.8 show the data flow for the complete environment. It is apparent that there is a very high requirement for interconnectivity of the test, repair and manufacturing systems. To date on the factory floor, different vendors have adopted proprietary communication architectures to consolidate their position. This is certainly true in the test environment. It is essential that the test environment integrates into the manufacturing and process control environment for full interworking. Presently it is necessary to adopt a 'least common

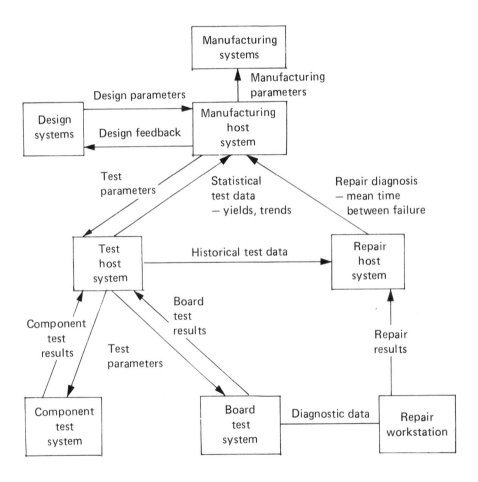

Figure 7.8 Data flow test/repair environment

denominator' approach to communications. However this often results in the use of RS232 communication between the system, which is relatively slow and has no inherent error detection and recovery. The use of RS232 LAN interfaces provide a degree of switching between the systems. The higher bandwidth of the LAN provides virtual multiplexing over the common cable, significantly reducing the cabling problem. However, this is not an ideal solution.

OSI standards, and MAP in particular, provide the infrastructure for high-speed resilient communications between the multi-vendor systems in the test and manufacturing environments. Moreover such a communications architecture allows the true integration of the test and manufacturing environments, without the need for expensive and inefficient gateways. Such an approach is infinitely more acceptable and realistic than using the 'least common denominator' RS232 method, or many expensive gateways between the proprietary networks.

Use of the wide area network WAN interfaces to MAP networks allow for the interconnection of many distributed manufacturing environments. This is particularly important for larger companies which may have a centre of excellence for design and development and a number of distributed manufacturing facilities including offshore facilities. The exchange of data between these sites is crucial for the efficient operation of the production process. Production statistics can be collected automatically from each site, acting as a feedback into the design system as well as overall monitoring of production yields at each site. Based on this information individual or global production parameters can be returned, so closing the feedback loop over a wide area network.

MAP has significant uses outside manufacturing industries. The first major use is as a broadband data communications network providing a high bandwidth for data communications allied with the cable TV aspects of broadband networks providing security and so on. The second major use is as a control network for energy management control systems or fire/security control systems within a complex. Therefore, MAP and broadband local area networks can provide the necessary communications infrastructure for today's 'intelligent buildings'. P-E Consulting Services, in association with Scott, Brownrigg and Turner are considering all these aspects and providing appropriate solutions within the 'WORKPLACE 2000' programme.

7.5 AIMS – An Assembly Information Management System applied to engine assembly

Introduction

The Cummins Aims System was devised in 1985 prior to the availability of either a suitable version of MAP and (as a consequence) an acceptable range of MAP compliant equipment.

It was however recognized by both Cummins and the participating suppliers that the role of standards and conformance to those standards expected to be included in MAP was essential to the long-term viability of the project.

In 1986 the AIMS system was re-implemented in a limited form for the CIMAP Exhibition, with both MAP and TOP conformance being achieved through the use of the appropriate software and gateway devices. Although this is not perceived as an economically desirable change to live system at the moment, it did show that there were no performance barriers to inhibit the use of MAP and or TOP in future projects.

The following case study is therefore included as being an example of the form that large scale integration takes and the need for a defined system architecture as well as MAP and TOP conformance.

Honeywell Bull and Cummins Engine Company have created what is believed to be one of the most advanced engine assembly plant in Europe. This completely new plant, at Darlington in England, turns the just-in-time (JIT) philosophy into a reality. The Honeywell Bull computer system AIMS (Assembly Information Management Systems) enables this plant to build engines to a wide range of specifications on a random basis with no changeover time. This section describes the computer system which allowed this project to be realized and demonstrates how computer integration can be applied to a very wide range of product manufacture.

In 1986 the Cummins Engine Company began production using their new engine assembly facility in Darlington (see Fig 7.9). The facility had been designed from scratch and targeted to meet the following objectives:

1. Flexible assembly of the Cummins 'B' series engine with three, four or six cylinders and accessories built to customer order, with single lot sizes.

2. Engine assembly in any order, at random down the line, with no changeover time.

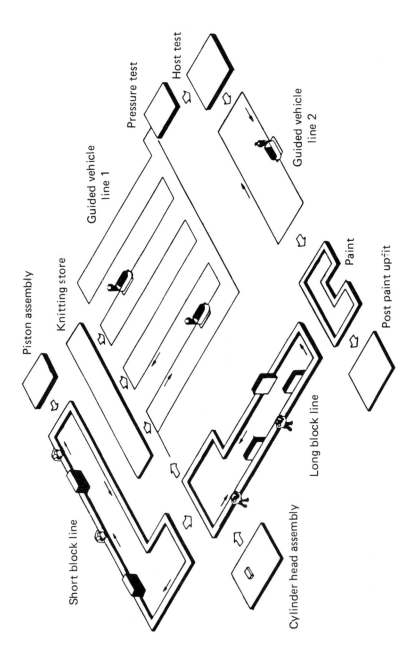

Figure 7.9 Schematic illustration of the Cummins engine assembly plant

3. Inventory turn improvement by having minimal stock at all stages of assembly, from material receipt to finished goods despatch.

4. Implementation of JIT techniques including qualified vendors, line side deliveries, no goods-in inspection, and no goods-receiving stores.

5. Optimal use of resources and real-time situations through decision support availability to supervision.

6. Team concepts for assembly personal, with each team having total responsibility for their section of the assembly line.

7. Quality audit, where test details and serial numbers of major components are captured and associated with each individual engine manufactured. thus providing every customer with a 'product of pedigree'.

To meet these objectives, Cummins divided the assembly line into a number of key areas and the equipment required for these areas was contracted out as turnkey elements.

The major items were a semi-automatic power and free conveyor line for the assembly of the base engines, principally supplied by Johann Krause, and two automated guided vehicle lines with 'ride-on' assembly platform; the AGVs were supplied by Jungheinrich. The two automatic guided vehicle lines were used for assembling the customer specified components and to move between the pressure text cells, the hot test cells and the paint system. The business systems computer used by the Darlington plant is IBM.

Having specified the assembly line completely, Cummins and Honeywell together then specified the requirements for AIMS. This system was to provide the integration between the IBM business system and all the process-connected devices.

The main functions AIMS had to fulfil were:

1. Mainframe interface

A systems network architecture gateway was required to interface to the MRP system and also to provide terminal emulation facilities from all AIMS business terminals.

2. Real-time work in process tracking

Cummins needed to know the status of all engines on the line when required.

3. Real-time machine status monitoring and reporting

Many sections of the line run completely automatically. Therefore, it

4. Provide all the assembly information required by machines and operators

AIMS interfaces to all the PLCs, AGV controllers, test cells and paint system to provide all required assembly information. Manual operations are provided with a parts list display which displays the parts to be fitted to the engine at that operation. This is updated as each new engine arrives.

5. Factory data collection

All test results are automatically gathered and stored for every engine. In addition, all of the major component serial numbers are captured from industrial terminals on the assembly line. At the end of each day, all engine history is passed on up to the MRP system.

6. Line supervision decision support

This is based on collected data such as machine availability and parts status which allows optimum use of a given set of circumstances.

7. Line schedule edit

The line set schedule is received from the MRP at the start of each day. This can be manually edited at any time up until the engine is committed to the line. In this manner 'rush orders' can be accommodated almost instantaneously.

8. Enquiries and reports

Cummins required the facility to be able to specify various enquiry and report formats to integrate the data held in AIMS.

9. Parts file edit

The parts files are received from the MRP business system. These can be edited locally if desired.

10. Picking list printing

For all customer specific components that are to be assembled to the engine on the AGV lines, a picking list is printed in advance in order that a kit can be made up for the engine. As the engine is loaded onto the AGV line, the kit is then put with the engine for the AGV.

11. Sub-assembly co-ordination

The co-ordination of sub-assemblies eg the cylinder head, for each engine is the responsibility of the AIMS system.

7.5.1 The solution

The Honeywell Bull solution was based on the company's

manufacturing automation architecture. This architecture recognizes the dynamics of information flow at the various levels ranging from sub-second response at the point of manufacture to a slower, though sometimes not much slower, response in other levels, and allows for the provision of the appropriate solution.

These levels can be described, starting from the top of the business as:

1. Enterprise level

This is concerned with corporate business decision support at which level activities such as financial modelling and analysis, business tracking and reporting, market research, new product planning and design, payroll and general ledger are common.

2. Factory level

Material requirements planning; labour and inventory control; production planning; scheduling and forecasting; these and others are the activities carried out at the factory level.

3. Supervisory level

At the supervisory level labour and facilities availability, quality control, maintenance management and scheduling, optimization of production are prevalent requirements.

4. Manufacturing cell level

It is necessary at this level to co-ordinate many different manufacturing capabilities including robots, programmable machine controllers, conveyors, guided vehicles, operators to blend these separate entities into a flexible efficient production unit.

5. Machine level

Direct control of machines and processes is carried out at this level involving sequence and logic control, manufacturing data capture, closed loop control, etc.

6. Sensor and actuator level

This is the area of parameter measurement by various sensing methods, and the man-machine interface.

Additionally, the cell level is implemented with 6800 and VME technologies, and the cell controller software is developed and tested in an advanced engineering environment. The programs are highly modular, written in 'C' and subject to automatic change control and audit. During operation, data reliability is maintained using secure communications protocols and bar code data capture from operators.

7.5.2 Autonomous operation

The principle of autonomous operation is achieved by distributing data throughout the systems. Data required to perform a particular function resides at the lowest appropriate level.

In the supervisory layer, a schedule for the operation of the line is maintained for three days' work. Extracts from this schedule are sent to the cell layer as advanced details for each engine. Each cell controller maintains a local database of this data for all the engines within its tracking boundaries. In the unlikely event of supervisory failure, the cell controller continues to operate using these advanced details and will store all engine history data for up to three shifts, for subsequent transmission to the supervisory computer when it recovers.

7.5.3 Real-time response

The cell controllers employ an operating system which allows them to handle the high rate of asynchronous transactions which are necessary to maintain control over the operations within the cell boundary.

Each monitored shop floor operation has a corresponding program, which remains dormant until activated by a communication from the operation's machine controller or from an operator. The six cell controllers have in excess of 150 such operations distributed between them, thereby modelling in the software the shop floor arrangement.

When an operation is activated by a communication it typically triggers a number of transfers of data between the machine, cell and supervisory levels. For example, when an engine arrives at a particular station, details regarding the engine are displayed to the operator. These details can originate from the cell controller if available within the controller's database, or if unavailable, can be acquired by the cell controller from the supervisory computer. Normally, the display of data regarding an engine is displayed instantaneously with the arrival of the engine at its station. However, should the information not be resident within the cell controller, then a typical time till display is two seconds, where data is acquired from the supervisory level.

The cell controllers currently operate with an average of 195 data transfers per minute, but are designed to allow in excess of 340 transfers per minute, and the volumes of data traffic range between two kilobytes for the complete details on an engine to ten bytes for tracking data.

A major real-time demand is the ability of each cell controller to service simultaneous communications between itself and multiple devices. A combined data rate in excess of 7,500 characters per second with protocol response times of 300 milliseconds was achieved, for multiple channels, using advanced multi-protocol communications devices.

7.5.4 Standardization
At peer levels in the hierarchy the hardware platforms and software environments are standardized. Each cell controller, while performing a different task is interchangeable with other cell controllers. This eases the problems of maintenance and upgrade, which can be implemented over remote links offsite, from the main facility.

7.5.5 The future
The AIMS concept provides a solution which is in itself flexible and appropriate to many diverse manufacturing areas. The techniques developed and proved at the Cummins plant are currently being successfully applied to businesses ranging from food production to the manufacture of electronic boards.

The Cummins project at Darlington has proved the effectiveness of integrating factory automation through the use of distributed computing.

7.6 Using MAP in the factory

Introduction
MAP installations have been implemented for three different factory applications and two MAP demonstrations by Hewlett-Packard's Advanced Manufacturing Systems Operation (AMSO) within the last two years. These implementations have demonstrated the feasibility of MAP and the concept that many vendors are able to join in a common effort to provide multi-vendor communications over a single broadband backbone network in the factory. The implementations to date have used MAP 2.0, MAP 2.1 (Autofact Subset), and full MAP 2.1. The manufacturing message formatting standard (MMFS) and the file transfer access and management (FTAM) provide peer-to-peer communication and file/program transfer capabilities to develop any type of monitoring and control application for the factory.

7.6.1 MAP implementation
Hewlett-Packard has been involved with the General Motors MAP standardization activities since February 1982, when GM asked for help from major computer vendors to develop a communications standard for the factory environment. Hewlett-Packard (HP), IBM, and DEC were the three primary computer vendors asked to review a draft

specification and to participate in the MAP Task Force Subcommittee meetings where the details of the specifications were worked out. Some of the key milestones in HP's MAP development are listed below:

June 1984 to
May 1985   *MAP 2.0 gateway developed and delivered*
- Provided access across MAP 2.0 broadband network to Allen-Bradley and Gould-Modicon PLCs as if they were attached directly to the MAP broadband; multi-vendor network of HP, DEC, Concord Data Systems, and 3M nodes; first MAP broadband pilot delivery to GM; HP defined MMFS PC level 0 specifications from GM action field definitions.

July 1984   *NCC '84 National Computer Conference Demo of MAP 1.0*
- HP was one of six vendors demonstrating multi-vendor connectivity over one broadband cable using MAP.

November 1985 Autofact '85 MAP 2.1 multi-vendor demonstration
- 23 vendors participated on three different sub-networks.
- HP provided job dispatch system which communicated with PCs and with all end nodes.
- HP's AMSO developed application and graphics software for the demo.
- HP's Roseville Networks Division (RND) staffed, developed, and delivered the MAP protocol stack for the demo in approximately 11 months.

January 1986   *First HP Cell Controller to major automotive manufacturer.*
- Multi-vendors: HP, DEC, INI, GMF Robotics, CMI Robotics A900 Cell Controller used the INI MAP 2.1 NIU (network interface unit).
- This is HP's first fully developed cell controller. It is expected to become a prototype for many more to be shipped to automotive plants.

Automotive manufacturer MAP pilot-initial delivery (March 1986)
- INI CASE Interface on A900 for MAP pilot cell controller development. Additional AMSO deliveries in April-June.
- HP is developing a fully functional MAP application I/F including a MMFS responder for this pilot.

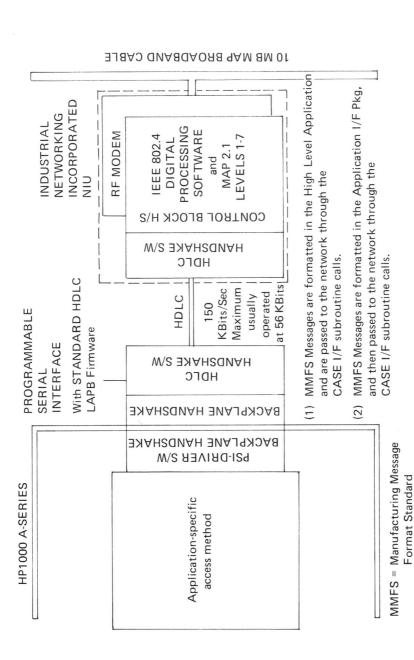

Figure 7.10 HP 1000 Network Connect using INI. For access method see Fig. 7.11

7.6.1.1 Industrial networking incorporated (INI) interface
This interface is the one being used for two major automotive manufacturers (Figure 7.10). All seven layers of the MAP 2.1 protocol stack are implemented in the INI box, and there is a control block handshake between the mainframe and the INI box. This control block handshake happens over a high-level data link control (HDLC) link that uses the standard programmable serial interface (PSI) card with standard HDLC LAP-B firmware. The standard PSI HDLC driver is also used, and the control block handshaking takes place in the INI CASE I/F module that also provides the buffering between the network and the application programs above it. Hewlett Packard's MAP Application Interface Package (AIP) software provides the standardized MAP interface to the network for all applications in the mainframe. In the two projects, this interface provides exactly the same functionality, but each pilot is using it differently in their own plants.

7.6.1.2 MMFS Robot Class 2
MMFS is the manufacturing message formatting standard. Robot Class 2 is the particular subset of MMFS used for MAP 2.1 and for these Pilots. It provides the formatting of the messages which are transmitted across the network through the MAP 2.1 protocol stack. The name 'Robot Class 2' is very misleading, because it is not particularly aimed at the control of robots. It is oriented more at the monitoring and control of intelligent devices versus the 'response only' devices usually found on the proprietary PLC networks. The two versions of MMFS that have been implemented to date are the PC Level 0 used in the HP Gateway and the Robot Class 2 used in the HP Cell Controller.

Typical MMFS messages and their use are shown in section 7.6.4.

7.6.2 MAP network access methods
Factory floor applications have many methods of access to the network, depending upon what type of communications is desired. Figure 7.11 shows the methods that an application can use to communicate with another application on a remote node or with a remote MMFS responder, such as the one provided by the HP AIP.

Application using CASE
Applications that will format their own messages or want to use non-standard message formats would access the network directly through the CASE interface services. This means that the application would have to manage the connections and respond to incoming indications from the network directly. This is not very difficult, since the CASE interface is very simple to use. Basic services provided are

MAP and TOP Application Case Studies 143

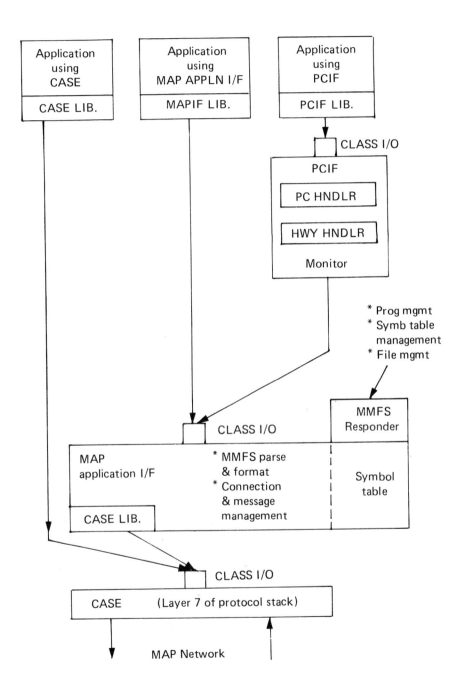

Figure 7.11 Methods of accessing the MAP network services

those to manage connections and to transmit and receive data.

Application using MAP Application I/F Package (AIP)
Those applications that want to use the MMFS formatted messages or use the more extensive services provided by AIP would use the AIP library access routines. This is the most likely interface method used by any application on the MAP nodes. The HP AIP provides all of the basic MAP services that are possible through the use of MMFS and the CASE I/F. The following services are provided:

- Message formatting, parsing, and routing
- MMFS message formatting and parsing
- Connection management
- File and program transfer (upload and download)
- Program management (select, start, and stop)
- MMFS responder capabilities for local and remote applications
- Symbol table access

Application using PCIF
The currently existing applications that use the HP standard product Programmable Controller Interface (PCIF) will have transparent access to devices that are attached to the MAP Network. There are also enhancements that have been added to the PCIF library to provide access to the additional capabilities provided by the AIP. These additions will allow an application to have a very sophisticated interaction with a MAP-connected device without having a complicated program structure. The PCIF interface allows an application to forget about protocols, connection management, and formatting messages for devices. This is all taken care of by PCIF.

7.6.3 Data flow and message traffic
Typical data flows and the relative traffic volume are shown in Figure 7.12. This figure is representative of several implementations, but the flows are so similar that it could represent almost any typical pilot MAP installation. It reflects the automotive industry and the types of installations and proposals discussed at AMSO over the last year or two in the evaluation of proposals and requests by their customers.

The messages flowing between the area manager and the cell controller are oriented towards background data, and program and historical data files necessary for statistical analysis and larger scale flow of materials. The cell controller typically should be able to operate when the area manager is down, and in installations where real-time process monitoring and control are desired, the cell controller is

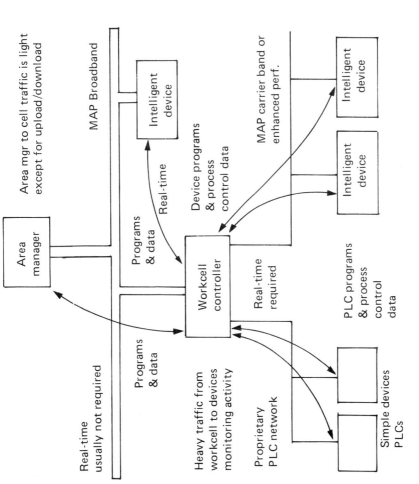

Figure 7.12 Typical data flows and message traffic

actually involved in the control of the production line or cycle. The cell controller can provide a 'pass-through' capability as well as a 'monitor and control' capability. The cell controller provided by AMSO for a major automotive project is used initially as a 'pass-through' workcell. It is expected that as the customer becomes more familiar with distributed control, the factory floor personnel who actually use the cell controller and control its configuration and operation will begin to use its real-time response capabilities more.

For example, the area manager will send a MMFS 'write' containing BUILD DATA option parameters to the Cell Controller as a 'write' to a named variable. This data is to be downloaded into each of the programmable controllers as a particular vehicle enters the workstation controlled by the PLC. A vision system's Smarteye barcode reader will detect the vehicle, read its VEHICLE ID number on the carrier, and send the information to the HP A900 cell controller. Monitor and control software within the cell determines that the vehicle is entering the new station, looks up the correct BUILD DATA in the BUILD DATA Buffer, and downloads it to the PLC. This all must happen very rapidly to ensure the PLC is ready when the vehicle enters the workstation.

The messages flowing between the cell controller and the factory floor devices are oriented towards the monitoring and control of the production process itself. The factory floor devices actually control this production equipment, and the cell controller provides the monitoring, control, and support of the factory floor devices. This includes program upload and download capabilities, register and logical name reads and writes, and error/alarm handling. MMFS is the message format used to communicate with all devices that are attached to any type of MAP communication media. GM pushed extremely hard to require that all their vendors provide MAP communications before submitting a response to proposal starting in January 1985.

7.6.4 MMFS Robot Class 2 – typical messages

Some typical MMFS messages used in the MAP Pilots are shown in Fig 7.13. These messages are a sample of what can be done between nodes and applications which are supplied by different vendors. These messages can be received by anything which is MAP 2.1 compatible. They will respond in a predictable way and take the appropriate action requested in the message.

These messages are used to share data across the network and to manage programs on a remote node. Notice that data names can be defined to be either local or remote and that they can be either read or

MAP and TOP Application Case Studies 147

```
                    MMFS Robot Class 2

Read data:
  REQ:  <OC:ui><TN:ui><PC:FRS><I2:REA><DN:ch>
  RSP:  <OC:ui><TN:ui><PC:FRS><I2:REA>{(single_element),(array)}

              (single_element) := <(type_id):data>
              (type_id) := {BS,SI,UD,BO,CB,UI,BD,CH,FP}
              (array) := <CT:ui>{(single_element)...,(data_stream)>}
              (Ddata_stream) := [<DF:(type_field)>]<DS:ui<<data>>

Write data:

  REQ:  <OC:ui><TN:ui><PC:{CMD,CRQ}><I2:WRI><DN:oh>{(single_element),(array)}
  RSP:  <OC:ui><TN:ui><PC:FRS><I2:WRI>

Program select:

  REQ:  <OC:ui><TN:ui><PC:{CMD,CRQ}><FO:SEL><CH:ch_data>
  RSP:  <OC:ui><TN:ui><PC:FRS><FO:SEL>


               MMFS Robot Class 2 functions

  Alarm          File-program      Identification    Program select
  Calibrate      Delete            Read data         Hold
  Reset start    Directory         Write data        Stop at end
  Continue       Upload            Status            End cycle hold
  Cycle start    Download                            Take control
                                                     Rel control
```

Figure 7.13 Sample of MMFS Robot Class 2 messages

written by a local or remote application. At this time, the data names used must be preset by the person(s) writing the applications on either side, in order to understand the meaning of the data. This is usually accomplished by publishing a naming convention for a particular installation or it can be done by specifying reserved names for commonly used data.

Remote program management is a very powerful capability which can be used for many things in the network. At the present time, the capability is restricted to one program on the remote node at a time. HP has recommended that the program name field be added to the start and stop MMFS commands to allow remote control of programs/processes in a multi-tasking operating system.

7.7 MAP in printed circuit board assembly

At the beginning of 1987, the UK's only mainframe computer company, ICL, began the second stage of its MAP project. ICL believes its Kidsgrove printed circuit board (PCB) factory will be the first UK factory to use MAP to link computers and work cells on the production floor and it planned to demonstrate the system by autumn 1987. The Department of Trade and Industry (DTI) has offered a 25 per cent grant towards the £500,000 ($800,000) first stage of the scheme. One significant result of the DTI grant is that it is only provided if ICL allows visitors to see the cells working. This means ICL Kidsgrove and British Aerospace's Warton site (see section 7.8) will be among the first UK factories to demonstrate cells linked up on MAP communications.

ICL intends to build the network in two phases. The first, fairly elementary, phase is to link two of the four levels into which ICL has divided its corporate communications hierarchy (Fig 7.14): these levels represent the plant hierarchy and should not be confused with the seven layer model. At the top of ICL's four levels is the 'OMAC' production, control, scheduling and materials requirements planning computer, an ICL 2900. This sends requests for materials down to level three, a System 25 computer which carries out stores management tasks. It sends picking lists to an ICL personal computer at level two which controls each of three Dexion vertical storage carousels at level one. All the components which go to the shop floor go through the carousels. Depending on the picking lists sent from level two to level three, the personal computer moves the carousel round to the operator's position and gives picking instructions. Levels one and two are linked by RS232 communications links. So are levels four and three. ICL believes MAP

MAP and TOP Application Case Studies 149

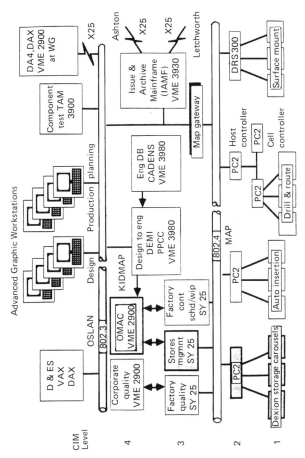

Figure 7.14 ICL communications hierarchy at their Kidsgrove PCB factory

is appropriate to link levels two and three. At first they will be linked over a MAP 2.1 broadband cable, later to be replaced by MAP 3.0, as soon as the new specification is released.

Phase one, the twelve-month planning and feasibility study was completed at the end of 1986. Phase two is the longer term spreading of MAP communications systems and controls throughout the shop floor. One of the first steps in phase two was to begin to arrange some of its warehouse and drilling machine operations into cells connected to a MAP communications backbone. This backbone will eventually link every stage of Kidsgrove design and production control procedures – from materials requirements planning through engineering design, automatic drilling and component assembly to final testing – on a single open-systems network which ICL is calling KIDMAP.

ICL also seeks to eliminate paper tape from programming of the drilling machines, which are used to make holes in its printed circuit boards. ICL wishes to set up a cell controller which sends the program to the right machine, cutting out the complexity of control hardware on each machine. ICL says that at each drilling machine it has had to provide a microprocessor-based box that looks like a paper tape to the machine: none of them had even so much as an RS232 port. ICL has now linked fifteen of these boxes to one personal computer.

Currently, designs are produced at a number of UK sites in various forms: drawings, tapes and discs. ICL's ultimate aim, according to David Thompson, is to eliminate all physical media from the company's design cycle and achieve a paperless shop floor. ICL says the KIDMAP network will allow data generated on CAD systems in design offices to be transmitted electronically to the assembly and test equipment in the factory. This will eliminate what ICL's Jim Kenny calls the 'log jam', the need for traditional media such as paper tape, cassettes and floppy discs to be carried about the shop floor from process to process.

Links to test equipment are important because of the need to produce precise test data. Linked CAD systems will be able to produce this test data at the same time as a board is designed. Every outgoing board is tested and there is also a great deal of time and effort devoted to goods-inwards testing.

ICL's installation is likely to prove interesting in two other respects. One is that it will seek to use part of the broadband coaxial cable for an Ethernet-type subnetwork – the latest versions of the MAP specification make allowance for such broadband-Ethernet networks. The second is that ICL intends to make full use of the network's large signal capacity to help operators, with consequent improvements in throughput and product quality. The way this will work is that, if an operator is unsure of his or her next operation, a signal can be sent to the cell controller which will then access an optical disc. A frame or sequence of frames will be selected from the optical disc to show the operator what to do next. These will be relayed to a TV screen next to the operator. Jim Kenny says this will be such an effective training aid that the extra production obtained from quicker-trained operators will more than outweigh the cost of preparing the video material the operators will use.

But ICL isn't investigating and installing MAP merely for its own internal reason: it wants to acquire enough MAP expertise to be able to develop and sell MAP interfaces for its own computers. Apart from the broadband cable, which is being installed by Ferranti (UK), the test equipment, the INI head end and interface boards and the vision-inspection system, all other items in the network will be developed and made by ICL. This includes gateways between ICL's 'OSLAN' TOP network and the MAP broadband. ICL will base its own installation on MAP 2.1 because, until 1988, that is all that will be available. But it has no intention of providing any products which conform to an earlier version of MAP than MAP 3.0.

ICL is particularly keen to develop its knowledge of and skills in the enhanced performance architecture (EPA) version of MAP. The EPA is a fast-response local version of MAP in which four – in some implementations, five – of the seven layers which make up the full MAP specifications are dispensed with (see section 4.8.3). ICL is working with Dextralog on an ESPRIT project to develop a cell control for a machine shop using the EPA.

The connections between the host and the MAP broadband cable in KIDMAP will be a full implementation of MAP. The automatic component-insertion machines and other equipment will be linked to MAP using a host or cell controller.

One of the things ICL has found most difficult is that none of the big suppliers of shop floor items for electronics factories has yet come up with or bought in carrier band interfaces for their equipment. ICL believes that suppliers of test equipment particularly, have an obligation to make carrier band interfacing cards available for users to retrofit to the large installed base these companies have built up.

ICL has set up a separate MAP development laboratory at its West Gorton, Manchester, site so that development work does not interrupt production at Kidsgrove. Kidsgrove is also being helped by the Engineering Services Division at West Gorton and Network Systems at Stevenage.

7.8 Communications in the aerospace industry

Some European companies are at least as far ahead as anyone in the USA in MAP implementation. One reason is the European Strategic Programme for Research and Development in Information Technology (ESPRIT). ESPRIT is in five parts: advanced microelectronics, software technology, office automation, advanced information processing and computer-integrated manufacturing. By the end of 1986 28 CIM projects were under way under six main headings: Integrated systems architectures; computer aided design and engineering; computer aided manufacturing; FMS; subsystems and components; CIM systems applications. The last provides the funding for CIM applications centres where contractors in ESPRIT can test their developments against those of other contractors. The commitment to CIM in entirety involves 150 contracting organizations – 30 of them small and medium-sized companies – contracting over 1500 man-years.

The key MAP project in the CIM part of ESPRIT is project 955, a UK-led effort to demonstrate a communications network for manufacturing applications (CNMA). The prime contractor and project manager for CNMA is British Aerospace. Three other user companies are involved – BMW of West Germany, Aeritalia of Italy, and Peugeot of France – and six vendor companies – Bull of France, GEC and ICL of the UK, Nixdorf and Siemens of West Germany and Olivetti of Italy. They are joined by French systems engineering company TITN and the West German Fraunhofer IITB Institut. The first public CNMA demonstration was held at the Hanover Spring Fair in April 1987, followed by demonstrations at British Aerospace's Preston and nearby Samlesbury factories, the Regensburg factory of West German carmaker BMW, and the Turin factory of Italian aerospace company Aeritalia.

Project 955 is important not just because it will familiarize Europeans with MAP technology but because it looks certain to be an instrument for developing and improving MAP itself. MAP still needs a lot of work and one aim of CNMA and a related project, project 688, is to enable European organizations to play a vigorous part in defining the form MAP will take. The EEC sees MAP as merely one part of a set of standards which need to be defined for the factory of the future. Project 688 seeks to define many of the rest, in step with MAP, and is being undertaken by another consortium including GEC and British Aerospace of the UK. This project was funded in the 1984 ESPRIT programme.

Many of 688's 19 members are European aerospace companies, though it also includes IBM Europe. Project 688, A European computer integrated manufacturing architecture, or 'Amice', is investigating areas which take MAP for granted and go beyond it. These include data exchange, databases, data management and applications protocols at the top of the seven layer model by which MAP is defined. A third project, called PRESTO, consists of user-groups which will compare the results of the two projects and rule accordingly.

Project 955 is funded with several million European Currency Units over three years. It was announced in September 1985 and work started on 2 January 1986. The brief for project 955 says: 'It is the project's goal to define, promote and provide a multi-vendor network

MAP and TOP Application Case Studies 153

environment for manufacturing applications. Such an environment must be defined from the end-users' point of view, taking into account the state of international standardization and de facto standards. It is a strategic purpose of this project to ease and accelerate the emergence of a common European position which will be the basis for real multi-vendor implementation in a global CIM approach.' In effect this means that CNMA will base its communications standards on version 3.0 of the MAP/TOP specification or an intercept strategy towards those specifications. The brief also says that 'Test-beds, prototypes, demonstrations and publication of implementation guides will conclude every step of the project.'

After the first demonstration at the Hanover Spring Fair, the demonstration moved to the British Aerospace sites at Preston and Samlesbury, UK. British Aerospace has a small-parts flexible manufacturing system at Warton in which research was conducted with GEC until their arrangement ended in November 1986.

The CNMA consortium had done enough work on MAP by the autumn of 1986 to expose some practical flaws even in the as-yet-unpublished MAP 3.0. In particular, CNMA will use different protocols at the top of the seven layer model on which MAP is based. MAP 2.1, as used at Autofact, used the manufacturing message format specification. CNMA is using draft five of the document that will become MMFS's replacement, RS511, now called the manufacturing message service (MMS) (see section 4.2.3). RS511 went into draft six at the beginning of 1987.

The CNMA Hanover demonstration (see Fig 7.15) uses some equipment later transferred to British Aerospace's small parts flexible manufacturing system (FMS) at Preston (see below). The Hanover demonstration, a subset of that FMS, combined a fully automated three-axis machining centre, an unmanned transport system, a six-axis robot and manual load and unload stations. These were linked to a Bull SPS7 minicomputer, a GEC Gem80 programmable logic controller (PLC), a GEC-supplied Num760 CNC controller, an Olivetti M44 minicomputer, a Nixdorf Targon 35 minicomputer, and an M70 minicomputer and Simatic PLC from Siemens. The network comprised a Siemens two-way router linking a five-megabyte IEEE802.4 carrier band MAP segment with a 10-megabyte IEEE802.3 LAN segment, and a GEC-supplied internetwork relay operating between the 10-megabyte IEEE802.3 LAN and the 10-megabyte 'true' MAP broadband LAN.

British Aerospace is hosting two demonstrations for CNMA. One is BAe's workshop in Samlesbury (Figures 7.16 and 7.17) for building leading and trailing edges for the wings of the A320 Airbus. The

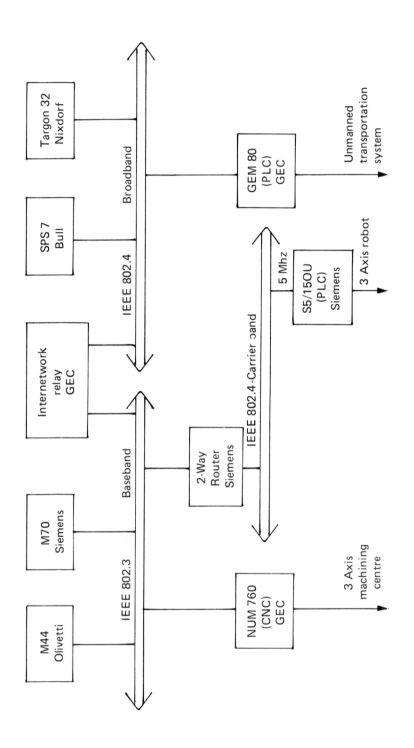

Figure 7.15 CNMA Hanover cell control architecture

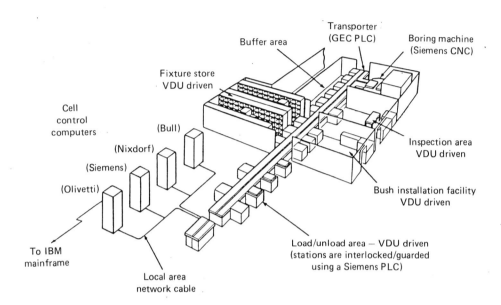

Figure 7.16 ESPRIT/CNMA demonstration A320 Airbus sub-assembly cell

assemblies for this are made up of many sub-assemblies which are similar in shape and function but differ in size because the cross-section of the wing alters in relation to the distance from the fuselage.

Each sub-assembly has parts which must be machined before they are mounted. This involves precision boring, inspection and bush installation. The transporting of parts around the cell and the precision boring operation are fully automated. Bush installation, inspection and loading and unloading of parts and fixtures, however, is done manually. Most factories will reach solutions which use a similar sensible compromise between automation and manual operations and the Samlesbury factory provides a good illustration of the contribution communications can make in a factory which has such a mix of work methods.

At Samlesbury the normal sequence is this:

- Procure empty fixture from fixture store

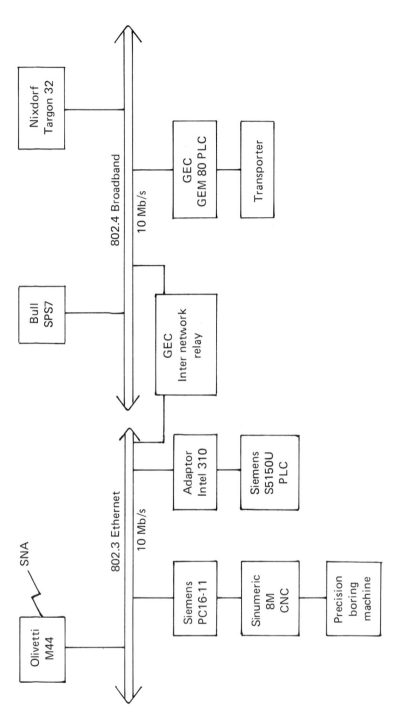

Figure 7.17 A320 Airbus CELL-control architecture

- Load parts on this fixture by hand
- Carry out first boring sequence on these parts
- Inspect bored parts
- Install bushes in these parts by hand
- Carry out second boring operation on the bushes
- Inspect again
- Unload the parts from the fixture by hand.

To achieve all this the cell controller needs to combine several functions. It must carry out production management and pure automation and it must interact with the production data processing systems elsewhere in British Aerospace.

The first task is 'job set-up', which means arriving at a 'job to do' definition. The definition identifies the parts, groups them up to a maximum of four to make the best use of the fixtures, defines the route the set of parts will take, and collects together the NC boring programs, the inspection programs and instructions and the bush-insertion instructions for the parts. The system has to make sure that all the tools needed for the job are available, along with the NC boring program, the bushing instructions and the inspection programs and instructions. This is all done by the cell co-ordination system. This actuates the automated subsystems – the conveyor or the NC boring machine, or gives prompts to the operator, whose instructions are displayed in a ruggedised VDU. Complementary functions are available to cell supervisors and to others, such as those in the stores or the subassembly area. The configuration is shown in Fig 7.17.

The Preston factory which houses BAe's small-parts FMS provides a second demonstration for CNMA (Figure 7.18). The Preston FMS is one of the most advanced of its kind in Europe, possibly the world. The FMS uses a database which holds all BAe's design, manufacturing and inspection data. This data is retrieved and disseminated on Ethernet LANs. Some of the links use fibre optic cable.

Communications are crucial to the Preston FMS because BAe's data management philosophy is not to hold data in the FMS. The data is stored in the supervisory system and down loaded to the FMS, which deletes the data after use. There is not even any local control: all the control signals the shop floor machines need come down to the shop floor from DEC supervisory computers. These in turn are linked via IBM's proprietary Systems Network Architecture communications system to BAe's IBM corporate mainframe. This 'drip-fed' approach to direct numerical control suits BAe's needs because Preston uses long machining times and very large volumes of data.

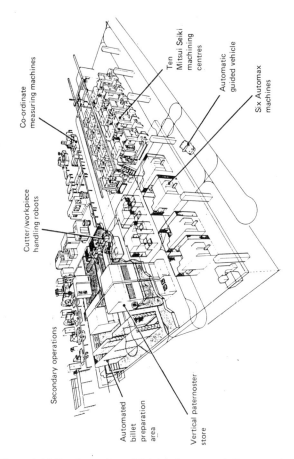

Figure 7.18 Configuration of British Aerospace's Preston plant

The CNMA implementation within the Preston FMS demonstrates three generations of communications (Figure 7.19): the all-proprietary fibre optic serial links between a DEC PDP11/24 computer and the machine controllers on four of the Automax machines; the use of an Ethernet (IEEE 802.3) LAN with proprietary protocols above layer two of the ISO model to link a DEC VAX and the Fanuc controllers on ten Mitsui machines for cutting steel and titanium; and the the CNMA profile, which uses ISO protocols from layers one to seven and links the NUM760 controllers on two of the Automax machines with the VAX through an ICL gateway. The ICL-to-VAX links use a preliminary version of FTAM.

The FMS, which will eventually work three shifts, is in two main parts. One makes components from aluminium. The other makes parts

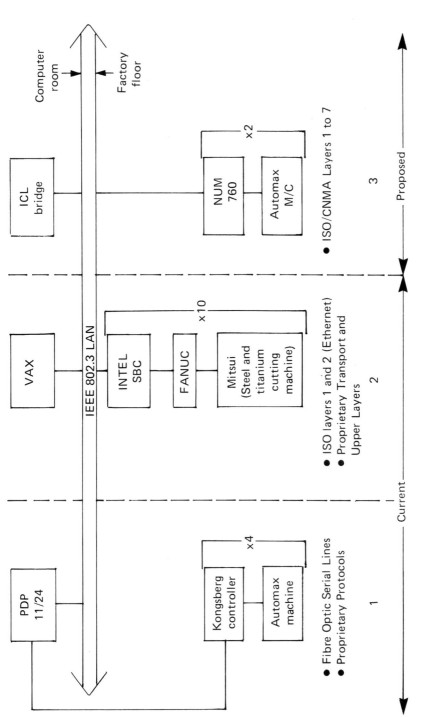

Figure 7.19 Production demonstration at British Aerospace, Preston

from steel or titanium. A third area, for cutter preparation, stores 80,000 tools of 3,000 different types for use in the FMS and the NC machines elsewhere in the Preston factory. The operation of this tool store also depends on an intricate system of links with the management and supervisory systems: even the tools are reground under DNC.

All the machined parts are made from aluminium or steel billets sawn from planks. To save material and cut machining times, several finished components are 'nested' together on one billet. The sawn billets go into a billet-preparation cell containing two modified Automax machines (see below) which face the billets on one side and drill and tap holes which will be used to attach the billets to fixtures. An order-picking robot stores all the prepared and raw billets on racking in the billet store. The racking can accommodate up to 400 raw or prepared billets, equivalent to about 48 hours' work.

The bar-coded billets are held against an order. On a signal from the supervisory computer, the order-picking robot delivers the billet to a conveyor and lift which takes the billet up to a mezzanine floor for fixturing. The fixturing process attaches the billet to a standard pallet covered in tapped holes one inch apart.

Part of the fixturing process is manual, part robotic. The fixturing robot has to know the fixturing patterns of each of the set of parts the FMS can produce. So far, the system has been programmed for 1,400 parts, though not all these programs have been proven. The robot also copes with empty pallets, which it unscrews for re-capping, and pallets with billets which have failed the inspection process, which it sends to a manual station.

Three vertical paternoster stores also act as buffers where the fixtured billets are kept until they can be machined. Each set of billets is stored along with a crate holding the set of tools needed to machine it. Each crate has locations for 64 tools: typically each job takes about 12 tools but some jobs can take as many as 31.

The crate of tools is assembled by a robot which picks the tool set from several hundred types of tool kept in a 4,000-location tool store next to the paternosters. The management system needs to know where each tool is and how much more life can be expected from it. All tools used for machining steel billets are reground after use so a lot of duplicate or 'sister' tooling is needed for these. The value of the cutters needed in the FMS is about £3 million ($4.5 million). The aluminium parts are machined on Automax machining centres. The Automax was developed by Marwin Production Machines of the UK with BAe, and is a two spindle machining centre with fully automatic workpiece pallet and tool crate loading.

After the billets have been machined on one side they are routed to one of two co-ordinate measuring machines for inspection following which they are loaded back into one of the bufferstores before going back to the mezzanine floor to be defixtured, turned over by robot and refixtured.

The FMS has two co-ordinate measuring machines. The first is an unmanned pass-or-fail station and the second is a diagnostic station which gives detailed information about parts that fail the first inspection. Both use part-programming data generated by the design system. The inspection process is tied closely to the management system. The management system knows how many times any particular tool has been reground and how many times it has been used since the last regrind.

The FMS has a full complement of eight Automaxes. One billet arrives at the mezzanine floor about every five minutes, though this will increase to one every three to four minutes. A steel billet arrives every half hour or so.

Steel and titanium parts are mounted on steel, cube-shaped fixtures. Two cubes are mounted at each of the ten Mitsui Seiki machines: one of the cubes is in the machine during the machining cycle; the other is outside the machine, at a station where finished parts can be removed from, or steel billets attached to, up to four of the cube's faces. Each cube can take up to ten or twelve hours' work.

The tool management of the steel part of the FMS comprises a robotic system installed by SI Handling of the UK. The cutters come from the bulk tool store to a small buffer store at one end of the FMS. A robot takes the crate of tools to the machine where they are needed and swaps the tools into the machine's tool carousel one at a time.

After the FMS the parts go through a crack-detection process, a heat treatment process, a part-marking process, then the parts are assembled into the aeroplane.

Tool preparation

As already mentioned, the cutter-preparation area stores 100,000 tools for use in the FMS and the NC machines elsewhere in the Preston factory. The tools are kept in a high density twin-aisle store supplied by Linvar. Two manned British Monorail semi-automatic rail-guided order pickers are used to collect and replace the tools from pull-out trays in the store. The order pickers are given picking instructions from a small terminal mounted on the order-picking car.

Tools are checked and reground in a regrinding area equipped with four eight-axis Huffman grinders which also operate under DNC. A cutter kitting and pre-set area prepares job sets of cutters. The three areas in the tool-management cell are controlled by a computer-based management system installed by CAP Reading. All tools, whether they arrive as new or come from the FMS on AGVs, are identified to the management control system. Used tools are removed from their chucks first. A bar-code label which identifies the tool is attached to its shank and it is inspected and passed to a regrind batching store. If the tool is reground the bar-code label is scanned again to reregister it and the file containing information about it is updated in the management system.

The tool is then put in a transport crate and put in the bulk store. A job schedule held in the management system puts together a list of jobs, their priority and which tools are needed to do them. When the cutter is taken out of the store for a particular job it is set accurately in a holder. The pre-setting data for each tool is recorded and fed to the DNC part program which controls the machine doing the job into which the tool is loaded.

Justification
The cost of the FMS is put at £10 million without the Mitsui Seiki machines. The project was justified on its own merits, albeit with a grant from the Department of Trade and Industry under the FMS scheme. That scheme awarded up to 20 per cent of the development and capital costs of eligible systems. BAe assumed that the project would not produce a return for four years and that the life of the machinery in the FMS would be 10 years. The parts made in the FMS now take about three days to make, compared with up to 12 weeks by traditional methods.

BAe's Brian Phillipson believes those involved in implementing communications systems should share their experiences much more. Minicomputer suppliers should share information with PLC suppliers, for example. He says there is even a lack of communication between different parts of the same organization, where one part is making PLCs and another is implementing them.

CNMA's practical approach is sure to have direct beneficial effects on the MAP programme, particularly since the consortium has set itself the task of developing conformance and interoperability tests for MAP equipment. These are two of the most difficult aspects of setting a single factory-communications standard.

A final point: those involved in CNMA feel sure that MAP and TOP are essential, but that they are not enough. They draw a comparison with ordinary conversation. Two people do not make conversation by delivering messages alternately in either direction. Conversation requires that each speaker understands what the other is saying – comprehension, in other words. A telephone will be able to deliver alternate messages effectively. It will not ensure that a Frenchman and an Englishman can converse.

The same is true of a minicomputer talking to another mini, to a programmable controller or to a CNC controller. MAP and TOP should be accompanied by the development of companion standards which ensure that, say, a Fanuc CNC controller understands the M and G codes on a tape in the same way as a Siemens controller understands it. There are many who agree with CNMA about this. There are even some who think progress in this area is much more important than standardizing the transmission medium used for communications.

The CNMA Implementation Guide is published by the Computer-Integrated Manufacturing task force of Esprit in Brussels, Belgium.

7.9 MAP in General Motors

As might be expected, no-one else has built up the MAP experience that General Motors has accumulated over the years. The following descriptions concern two of GM's most important projects: the Vanguard 'Factory of the Future' in Saginaw, Michigan (Figure 7.20), and the GMT400 Truck and Bus installation. The information about Saginaw was gained on a visit to the plant, the GMT400 description is based on conversations with engineers who have worked on the project.

7.9.1 Saginaw

It will be some time before GM will know whether its experimental factory of the future in Saginaw is a manufacturing milestone or an awful waste of millions of dollars. The Vanguard project at Saginaw is the first – and smallest – of three fully-automated factories GM is to build in the USA in the years to 1990. Vanguard's cost is usually put at $52 million. But this figure covers only the machinery being installed. It does not cover software and other support from GM's technical centre in Warren, Michigan, from GM's Electronic Data Systems subsidiary, from the Industrial Technology Institute in Ann Arbor, Michigan, from the University of Michigan, or from the suppliers to Vanguard, notably Stratus Computers and a small, San Francisco-based start-up called Maxitron.

The Saginaw Steering Division is one of 30 GM divisions. SSD has seven plants in the Saginaw area, eight in Detroit, three in Alabama, two in New York state, one in Spain and one in the UK.

The Vanguard project, housed in a 70,000 sq ft corner of Saginaw's

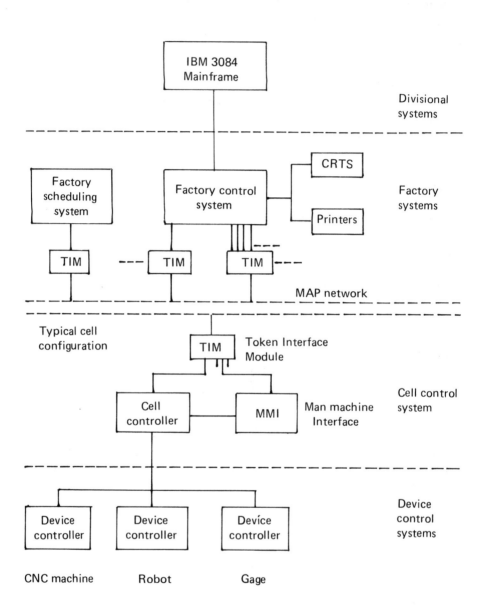

Figure 7.20 The Vanguard 'Factory of the Future' network at General Motor's Saginaw plant

plant four, is to make about 10 per cent of SSD's output of front-wheel-drive axles. This is a crucial feature of the project because the fact that Vanguard will make only 10 per cent of output will allow direct comparisons of cost, quality and delivery between parts made in the factory of the future and parts made using more conventional techniques. Each axle has 23 parts. When Vanguard is completed it will machine three of these parts – the housing, axle-bar and universal-joint 'spider' – and assemble these with the other 20 brought-in parts to make a finished axle. If Vanguard works, it will do all this completely unmanned under the control of the plant's CADCAM system.

The plant, conceived at GM's technical centre during 1982 and 1983 and approved in August 1983, has been designed to produce much larger components than it does now – the front-wheel-drive axle is less than two feet long and the largest components fit comfortably in a one-foot-cube envelope. In the start-up phase the factory is making only two sizes of axle but GM says it will be able to make either smaller or longer axles. The batch size is theoretically one but, because two axles are needed in each vehicle, in practice it is two.

Vanguard was due for completion during 1988. But by 1987 work was already far advanced on its 23 manufacturing and 17 assembly cells. Seven or eight of the manufacturing cells came on line in early 1987 to machine the housings and the spiders. The axles bars began coming out a few months later and assembly was due to begin after that. All the cells were to be linked up on a MAP network by the end of 1987. The link from CADCAM to the production-control system will be in place during 1988.

Despite being 'unmanned', the plant needs about 40 attendants divided among three shifts. High staff numbers are undoubtedly needed during the start-up phase but, people are still needed to load trays of components before the trays take the components into the cells. There are also six multi-function bowl feeders in the assembly area which must be loaded with the 22 part-types they use. Set-up sheets are all on paper. Maintenance documentation is contained in a 30-page binder, continually updated, which contains Polaroid pictures of machine-controller-VDU displays taken under particular fault conditions.

The control of tool and part movements and of model changes is carried out by a Stratus factory-control computer linked to a Maxitron cell controller in each of the cells. Maxitron has links to the French industrial-controls company Telemecanique. The Stratus is connected to a MAP network by a Concord Communications token interface module, or 'TIM box'. A TIM box linked to the MAP network exchanges data with each of the cell controllers. Within cells, RS232 links are used. The factory control computer uses a schedule held in another Stratus computer which is running custom materials requirements planning software. This second Stratus is also linked over the MAP network through a TIM box. The MAP network runs a continually-upgraded but still non-standard variant of MAP 2.0.

Changeover times between parts are stated as 'under 10 minutes'. Robots of various types are used to change chucks as well as load parts. If the chuck requires the use of a different end-effector, the robot will change its end-effector too. Tools are carried round by automatically-guided vehicles. AGVs are also used to collect and replace swarf bins placed at each machine. The swarf or chips, reach the bin by conveyor. Software-driven in-process gauging is used extensively. As the robot loads the tool into the machine, the tool is probed so that the robot knows how far to load it into the machine and the controller knows where the cutting point of the tool is. This information is used to amass cumulative information for quality control as well as to modify the controller software. As well as in-process gauging Vanguard uses a laser-shadow gauge which takes five measurements from the spider in two minutes. Vanguard's grinding machine, a Byant Lectraflex, does all the grinding on a spider – including inside and outside diameters – in one operation. Conventional methods, say those at Vanguard, would need six machines.

The assembly operation will, if it succeeds, contain some interesting features. One is that it will make use of pairs of robots equipped with vision systems which will assemble axle parts in free space. The cameras will be mounted either on the robot arms or on fixtures nearby.

The parts to be assembled are all collected at one end of the assembly area and conveyed to the appropriate assembly cells by a gantry-mounted robot. The robot is given the information it needs about the locations it must collect to or deliver from by a transponder on the pallets going through the assembly process. This is needed because the number of possible assembly combinations is so large that no local controller could accommodate the number of programs required to represent them.

When in full operation, Vanguard will have capacity to make 1,450 assemblies a day in three shifts. SSD intends it to reach full production by the end of 1987 but, if duplicate cells are added, this could be increased to 2,000 components a day.

One of the interesting things about the Vanguard project is that, despite the use of MAP, the Saginaw plant has standardized on one type of machine controller, the General Electric (USA) GE2000. However, the factory has a wider mix of robots, nearly a dozen, with different types of controller. This is an illustration of the way that GM approaches its communications needs. It wants broadband communications at a plant-wide level, but within particular production areas it will standardize as much as possible, providing proprietary subnets which link groups of devices through a gateway to the main MAP backbone. Saginaw is one example of this. The automation of GM's truck and bus group is another (see below). There GM standardized on Allen-Bradley controllers for fabrication plants and Gould programmable controllers for assembly operations.

It has been extensively argued, particularly by those involved in the CNMA project, that most companies are not as large as GM, and their production profiles are different too. It is this that has led to the development of the carrier band idea, wherein the elements used on the small subnets are also provided with uniform interfaces, linked over a cheaper but non-proprietary sub-network.

7.9.2 The GMT400 Truck and Bus Project

Vanguard is experimental and, because its approach to communications has been somewhat pragmatic, the version of MAP it uses is non-standard. For these reasons, General Motors' GMT400 truck and bus project is perhaps a better example of the day to day use of communications technology. The automation programme for the truck and bus group of General Motors covers five plants, all of which are dependent on the MAP networks installed in them: if the MAP network fails, the plant stops.

The truck and bus division is much bigger than the Saginaw project, which is a pilot. The five plants are Fort Wayne, Indiana; Pontiac, Michigan; Oshawa, Ontario; Flint, Michigan and Indianapolis. The last two are fabrication plants and the first three are assembly plants. The plants are due to make the vehicles for the 1988 model year.

Much of the task of installing MAP for GMT400 consisted in developing and writing interfaces. Much of the MAP seven layer profile was not available when GM subsidiary Electronic Data Systems began work on GMT400. Two of the main pieces of software which were needed were 'Autoway', a piece of software in the top layer which allows the area manager applications to have access to the MAP channel, and the cell controller communications integrated interface. There were no MAP 2.1 hardware interfaces for the DEC Vax 8600 computers which acted as area managers so these had to be developed with INI and a West Coast company called Cimtact. Interfaces were also needed for the IBM plant hosts – which hold enough data to keep each plant running for a week – and for the Hewlett-Packard cell controllers. The area managers hold eight hours' data and the cell controllers two hours'.

EDS also had to develop tests – again none had so far been written – for conformance and functionality: there was no need for interoperability tests because all the interfaces used were those of INI. Performance tests were also needed to see if the systems could support the overall plant architecture. The Autoway software had to be tested rigorously and the seven layer stack had to be tested layer by layer: the timings in layer four, where ten simultaneous connections were possible in one application (see section 4.5.1), proved particularly tricky. Other tests involved taking devices on and off the network at random times to see what the effect would be on the network.

If all this interfacing was needed, what was the point of the MAP network? Those involved say that MAP gives them the widest possible choice of new devices. When a new product comes along they can use it on the MAP network much more easily than otherwise. The robots used were GMF and Cincinnati but these are not interfaced directly to the MAP network yet; they work on subnetworks. Their programs are stored on cassette tapes but GM plans to download program changes over MAP into bubble memory in each robot controller. GM also wants to be able to pull error conditions back up over the network. The first assembly plant started in March 1986 and by the beginning of 1987 this plant, in Fort Wayne, was making 150 trucks a day in one shift. The final ramp-up rate is 60 trucks an hour or 960 trucks in two shifts.

Bearing in mind that EDS was new to manufacturing systems when it started GMT400, the project has gone well. GMT400 managers say the last actual MAP-related problem occurred back at the beginning of 1986. The coaxial cable also carries point-to-point communications for test-system data and video pictures from TV cameras. Those involved in the project say the fear of head ends failing is negligible. One manager notes that he has only experienced one head end failure since GM started using broadband in 1970 or 1971. This he attributes to the fact that MAP uses tried and tested cable television technology. However, there have been problems with modem drift – the gradual tendency of all electronic systems to wander out of tolerance as they come into operation or temperatures change. The last one these insiders could report, though, was four or five months earlier. The communications system is on all the time, even during the maintenance shift.

Those involved in this project did have experience of shutting the whole network down and starting it up again, and one facet of this is worth noting. One GMT400 manager, whose plant has about 25 cell controllers and about 35 MAP nodes, says that, unlike those involved in the UK's CIMAP demonstration, they did not need to bring each node on to the network one by one: at GMT400 they just switched it on again, regenerated the token and off the network went. This may be because EDS has provided automated network management for GMT400. None of the MAP specifications before the 2.X series had network management and even MAP 2.1's network management system is fairly basic (see section 4.2.4.3).

8
THE WAY AHEAD

8.1 The future

In a way it is pointless to speculate too much on what is likely to develop from the two main initiatives which this book has been concerned with. So much, after all, of MAP's future depends on the success of Western manufacturing industry, and that is a study in itself. TOP, though off to a slow start, may turn out to be the answer to a lot of the communications problems in off-the-floor trading in the world's financial markets, for example. The provision of standard, plug-in communications based on open systems could transform the prospects for cashless shopping in high streets in all but the smallest towns.

Immense problems have to be solved, not least those in conformance, interoperability and performance testing, before genuinely open systems can be taken for granted. When they are, no doubt new products and services will flow from the availability of open systems that can only be guessed at now. But whatever benefits can be gained from open systems, they can only be gained if international agreement is reached to make the systems genuinely open. The question then arises, what happens if an international agreement closes off an option that might be of benefit in a niche market? What if the 'standards overhead' makes a product over-engineered or too expensive? Does the standards community tolerate the proliferation of exemptions or 'near-standard' products? Will these need to meet minimum interoperability requirements with truly standard products, or should the market be allowed to judge?

There are two main points which users should bear in mind about likely developments. One is the likely direction OSI-based specifications will take. There are a number of possibilities here. One direction, which is being urged in some parts of Europe, is to make certain seven layer profiles of standards, such as MAP or TOP, themselves standards. If this happened it would probably be a retrograde step. It would mean that one of the crucial advantages of the seven layer model – the ability to update a single layer without affecting any of the others – would be lost. If MAP and TOP themselves became standards it would become much harder to add to them and update them. A related possibility is that multi-layer profiles could be standardized at layers three to six but users could have other options at the top and bottom of the stack. This resembles the differences that exist between MAP and TOP now.

The second important issue is this. None of the things discussed in this book means anything unless the applications which run on these systems meet the market need. For example, even when all the communications standards are settled, will a user who wants to install a direct numerical control system covering a number of machines continue to have to provide interfaces which affect a different post-process conversion for every single different type of controller on every different machine tool? How long will it be before there is a single, cheap and simple method of converting design data derived on one maker's CADCAM system to run on another maker's CADCAM system? Will TOP users one day get a standard method of exchanging spreadsheets between different systems? In the end, the answers to these questions will be decided by the users, if they choose to keep up the pressure.

8.2 MAP and TOP products

A large amount of MAP and TOP communications hardware has already come on to the market. In the case of the TOP specification, using as it does the relatively common Ethernet-type lower two layers of the seven layer stack, this is hardly surprising. But the MAP specification uses media access and modulation methods many of which have been developed specially for factory communications. The MAP specification is therefore perhaps the more challenging.

The table which follows is a list compiled, with one or two exceptions, from telephone interviews. It shows that many companies are supplying MAP 2.1 hardware and software and there are even some which claim to be supplying MAP 3.0. Some of these companies have thereby registered their intention to supply upgrades to MAP 3.0, when it is published. Others will not supply any MAP 2.1 products but will begin to supply MAP 3.0 within a short period of its specification being published. This list was updated as we went to press.

FACTORY COMMUNICATIONS – VENDORS/PRODUCTS

Supplier	Cables			LAN Hardware			Interfaces		LAN Connections		Gate Ways	Software		WAN			MAP/TOP Services			Standards	
	Fibre Optic	Base-band	Broad-band	Head Ends	RF Modems	Trans-ceivers	Stand Alone	System Boards	Routers	Bridges		Net Mgmt	Application	Other	VADS	X.25	Net Mgmt	Integration	Consult	TOP Vers	MAP Vers
Allen-Bradley		*	*	*	*	*	*	*													2.1
Apollo Computers		*	*																		3.0
ASEA Kabel	*																				2.1
BICC Data Networks	*	*																			3.0
Bridge Communications		*																			
CAP											*	*			*	*	*	*	*	1.0	2.1
Concord			*	*	*	*	*	*	*			*	*				*		*	1.0	2.1
DEC	*	*		*			*	*		*	*	*	*		*	*	*		*	1.0	2.2
EDS												*	*				*	*	*	1.0	2.1
ERA Technology																			*	1.0	2.1
Ferranti plc				**	*		*	*	*	*	*	*	*	*		*	*		*		2.2
GEC Ind Controls	*				*		*	*	*	*	*	*	*	*	*	*	*	*	*	1.0	2.1
Gould Electronics				*		*	*	*		*		*	*	*			*		*	1.0	2.1
Hewlett-Packard	*		*	*	*	*		*	*	*		*	*	*		*	*	*	*	1.0	2.1
Honeywell	*		*				*	*		*		*	*			*			*	1.0	2.1
IBM	*	*	*	*	*	*	*	*	*	*	*	*	*			*	*	*	*	1.0	2.1
ICL	*	*	*			*		*					*				*		*	1.0	2.1
INI/U-B		*	*	*	*	*	*	*	*	*	*	*	*			*			*		2.1
Intel Corp				*	*	*		*	*	*	*	*	*			*	*				2.1
ISTEL															*	*	*	*	*		2.2
ITL	*		*					*												1.0	2.2
Motorola				*	*	*		*	*		*	*	*								3.0
Olivetti		*	*		*			*		*	*	*	*	*	*	*			*	1.0	2.1
Prime Computer	*	*	*	*	*	*		*		*	*	*	*	*		*	*	*	*	1.0	2.2
Racal-Milgo																					
Siemens		*	*	*				*	*	*	*	*	*			*	*	*	*	1.0	3.0
Stratus Computer		*	*		*														*		2.1
Sun Microsystems		*	*																	1.0	2.1
Sytek Inc			*	*	*													*	*	1.0	2.1
Tandem Computers																					
Telematics							**									*			*		
Wang		*																		1.0	

USEFUL ADDRESSES

(ANSI) American National Standards Institute
1430 Broadway
New York
NY 10018
USA
Tel: (212) 3543300

Australian MAP Users Group
Bob Lions
Standards Association of Australia
PO Box 458
North Sydney
2060
Australia

CIM-OSA/AMICE
489 Avenue Louise
B14-B-1050, Brussels
Belgium
Tel: (32) 2 647 31 75

COM Centre
PERA
Melton Mowbray
Leicestershire LE13 OPB
UK
Tel: (0664) 501501

COS (Corporation for Open System)
1750, Old Meadow Road,
Suite 400,
McClean
Virginia 22102-4306
USA
Tel (703) 848 2100

DTI (Department of Trade and Industry)
Ashdown House
123 Victoria Street
London SW1E 6RB
UK
Tel: (01) 212 8675

EMUG (European MAP Users Group)
Cranfield Robotics and Automation Group (CRAG)
College of Manufacturing
Building 30
Cranfield Institute of Technology
Cranfield
Bedford MK43 OAL
UK
Tel: (0234) 752794

ESPRIT CIM
ITTTF DG XIII Commission of the European Communities
25 Rue Archimede
B-1049 Brussels
Belgium
Tel: (32) 2 2351111

General Motors MAP Information Centre
Tel: (USA) (313) 947 0555

FhG (Fraunhofer-Institut fur Informations und Datenverarbeltung)
Sebastian-Kneipp Strasse 12-14
D-7500 Karlsruhe 1
West Germany
Tel: (0721) 60 91

INTAP (Interoperability Technology Association for Information Processing)
Akasaka 7th Building 6 F
7-10-20 Akasaka
Minato-ku
Tokyo 107
Japan
Tel: (03) 505 6681

IT Standards Unit
Department of Trade and Industry
Room 634
29 Bressenden Place
London SW1E 6SJ
UK
Tel: (01) 213 5435

Japanese Industrial Standards Committee
c/o Standards Department
Agency of Industrial Science and Technology
Ministry of International Trade and Industry
1-3-1 Kasumigaseki
Chiyoda-ku
Tokyo 100
Japan
Tel: (813) 501 92 95/6

JMIT (Japan MAP Group)
Tado Tamura, IROFA
7th Floor, Daiichi Nau-Oh Building
2-21-2 Nishi Shimeashi
Minato-Ko
Tokyo 105
Japan
Tel: (813) 4332441

MAP Chairman
General Motors Technical Centre
Manufacturing Building A/MD-39
30300 Mount Road
Warren
MI 48090-9040
USA
Tel: (313) 5565000

MAP Secretary
Robin Haighton
The Canadian MAP Interest Group
c/o Canadian Standards Association
178 Rexdale Blvd
Rexdale
Ontario
Canada M9W 1R3
Tel: (416) 747 4149
Canada

The Networking Centre
Focus 31
Mark Road
Hemel Hempstead
Hertfordshire HP2 7BW
UK
Tel: (0442) 217611

The OSITOP European Users Group
(Open Systems Interconnection/Technical Office Protocol)
Mission Informatique et Telecommunications
21 Avenue de Messine
75008-Paris
France
Tel: (1) 47 64 27 94

POSI (Promoting Conference for OSI)
Kikai-Shinko Building
Room No 313
5-8 3-chrome Shibakoen
Minato-ku
Tokyo 105
Japan
Tel: (813) 433 1941

SMUG (Swedish MAP Users Group)
Sveriges Mekanstandardisering (SMS)
Box 5395
S-102 46 Stockholm
Sweden
Tel: (08) 783 82 92

Society of Manufacturing Engineers:
See US MAP Users Group

TOP Chairman
Boeing Computer Services
Network Services Group
PO Box 24346
Mail Stop 7C-16
Seattle
WA 98124-0346
USA
Tel: (206) 763 5973/(206) 865 6470

US MAP/TOP Users Group and World Federation of MAP User Groups
Patricia Van Doren
1 SME Drive
PO Box 930
Dearborn
MI 48121
USA
Tel: (313) 271 1500 ext 521

ABBREVIATIONS

10Base5	Standard IEEE 802.3 CSMA/CD 50-Ohm shielded coaxial cable. 10Mbps, 500-meter segments
ABM	Asynchronous balanced mode (HDLC Specific)
ACSE	Association control service element
ADCCP	Advanced data communications control procedures. The ANSI version of bit-oriented protocol; see HDLC
ADCP	Application dependent convergence protocol
ADM	Asynchronous disconnect mode
AFI	Address format identifier
AGV	Automatic guided vehicle
AI	Artificial intelligence
AIMS	Assembly Information Management System (Honeywell)
AIP	MAP Application Interface Package (Hewlett-Packard)
AMD	Advanced micro devices (US)
AM/PSK	Amplitude modulated phase shift keying
AMSO	Advanced Manufacturing Systems Operations (Hewlett-Packard)
AMT	Advanced manufacturing technology
ANSI	American National Standards Institute
ARCNET	Attached Resource Computer Network (Datapoint Corporation)
ASCII	American Standard Code for Information Interchange
ASN.1	ISO Abstract Syntax Notation one
ATE	Automatic test equipment
AT&T	American Telephone and Telegraph
BAS	Basic activity subset part of ISO session
BCS	Basic combined subset (ISO Session Layer)
BISYNC	Binary synchronous
BIU	Bus interface unit
BNA	Boeing Network Architecture
BPS	Bits per second (Transfer rate of a serial transmission)
BSI	British Standards Institute
BSS	Basic synchronized subset part of ISO session
CAD	Computer aided design
CAM	Computer aided manufacture
CASE	Common aplication service elements
CATV	Community antenna television
CBEMA	Computer & Business Equipment Manufacturers Association (US)
CBM	Carrier band modem
CCI	Concord Communications (USA)
CCIA	Computer and Communications Industry Association
CCITT	Comité Consultatif Internationale de Télégraphie et Téléphonique.
CCR	Commitment, concurrency and recovery
CD	Collision detection
CEN	European Committee on Norms
CENELEC	European Committee on Electrical Norms
CEPT	European Conference of Postal and Telecommunications Administrations

CGM	Computer graphics metafile
CIM	Computer integrated manufacture
CIM-OSA	European Computer Integrated Manufacturing Open Systems Architecture (ESPRIT Project 688).
CIMAP	Computer integrated manufacturing automation protocol (UK)
CIU	Communications interface unit
CLIP	Connectionless Internet protocol
CLNP	Connectionless network protocol
CLNS	Connectionless-mode network service protocol (ISO 8473)
CMM	Co-ordinate measuring machine
CNC	Computer numerical control
CNMA	Communications network for manufacturing applications
CONS	Connection-oriented network service
COS	Corporation for Open Systems (USA).
CRC	Cyclic redundancy check
CSA	Client Service Agent
CSMA/CA	Carrier sense multible access/collision avoidance
CSMA/CD	Carrier sense multiple access with collision detection
DA	Destination address
DAD	Draft amendment document
DAD2	Data communications network service definition – addendum 2
DARPA	Defense Advanced Research Projects Agency
DBX	Digital branch exchange
DCE	Data circuit-terminating equipment
DEC	Digital Equipment Corporation
DECnet	Trademark for Digital Equipment Corporation's communications network architecture which permits interconnection of multiple DEC computers.
DIN	German Standards Authority
DIS	Draft International Standard
DLC	Datalink control
DMA	Direct memory access
DNC	Direct numerical control
D&NC	Database & network communications
DOCMP	Digital data communications message protocol
DoD	Department of Defense (USA)
DOS	Disc operating system
DP	Draft proposal, ISO standard status
DP1	First ballot of the draft proposal
DP2	Second ballot of the draft proposal
DTE	Data terminal equipment – The equipment comprising both a data source and a data link with respect to a communications network. Typical DTE's are computer systems and/or terminals.
DTI	Department of Trade and Industry (UK)
EBQ	Economic batch quantity
EC	European Communities
ECL	Emitter coupled logic
ECMA	European Computer Manufacturers Association
ECSA	Exchange Carriers Standards Association
ECU	European Currency Units
EDS	Electronic data systems
EIA	Electronics Industries Association (US)
EMUG	European MAP Users Group Secretariat (Cranfield, UK)

ENV	European norms (Draft)
EPA	Enhanced Performance Architecture
ESPRIT	European Strategic Programme for Research and Development in Information Technology (sponsored by the Commission of the European Communities)
EXL	Fiberway Ethernet Accelerator
FADU	File access data unit
FCC	Federal Communications Commission (USA).
FCS	Frame check sequence
FDDI	Fibre distributed data interface
FDM	Frequency division multiplexing
FIP	Factory instrumentation protocol
FIPS	Federal Information Processing Standards
FMS	Flexible manufacturing systems
FSK	Frequency shift keying
FTAM	File transfer access and management
FTP	File transfer protocol
GCC	Gould Cell Controller
GM	General Motors Inc
GMF	General Motors Fanuc
HDLC	High-level Data Link Control. An ISO version of a bit-oriented synchronous protocol. Functionally identical to ADCCP (Link Level Protocol)
IAPP	Industrial automation planning panel
ICAM	Integrated computer automated manufacturing
IDC	International Data Corporation
IDT	Industrial data terminal
IEC	International Electrotechnical Commission
IEE	Institution of Electrical Engineers (UK)
IEEE	Institute of Electrical and Electronic Engineers (US)
IEEE 802	Activity to establish standards for LAN (Area Network Standard Series). Project 802 is divided into six Working Groups, which produce the standards, and two Technical Advisory Groups (TAGs) which advise the working groups. These are as follows: 802.1 Higher Layer Interface, concerned with the interface between the IEEE standards and the higher layers of the ISO model; also concerned with network management. 802.2 Local Link Control Protocol (network layer services) 802.3 CSMA/CD Medium Access Control Protocol (Ethernet based access method) 802.4 Token Bus Using Broadband Medium Protocol 802.5 Token Ring Using Twisted Pair Protocol 802.6 Metropolitan Area Network Protocol (citywide) 802.7 Broadband TAG is particularly involved with the 802.4 group. 802.8 Fibre Optic TAG is involved with groups 802.3 and 802.5, although it could apply to all the other groups
I/F	Interface
IGES	International graphics exchange specification
IH	Internet header
IISS	Integrated information support systems
ILD	Injection laser diode
INI	Industrial Networking Incorporated
I/O	Input/Output. The process of transmitting data to and from a computer; the data, devices or media involved in the process.

IOS	Intermediate open systems (or routers)
IP	Internet protocol
IPDU	Internet protocol data unit
IPL	Initial program load
IPM	Interpersonal messaging
IS	International Standard
ISA	Instrument Society of America
ISDN	Integrated services digital network
ISI	Interactive Systems Inc – now Allen-Bradley Communications Division
ISO	International Standards Organization
ISO/OSI	International Standards for Open Systems Interconnection
ITI	International Technology Institute (US)
ITSTC	Information Technology Steering Committee (Brussels)
ITU	International Telecommunications Union
IUT	Implementation under test
JIT	Just-in-time
JTM	Job transfer and manipulation
JTMP	Job transfer and manipulation protocol
LAN	Local area network (A LAN Manager is a program or system that manages layers 1 and 2 of a Local Area Network)
LAN-WAN	Local area network to wide area network communications
LAPB	Link access protocol balanced
LDIB	Local directory information base
LED	Light emitting diode
LLC	Logical Link Control
LPDU	Link layer protocol data unit
LRC	Longitudinal redundancy Checksum
LSAP	Link layer service access point
LSDU	Link layer service data unit
MAC	Sublayer Media Access Control layer, governs access to the transmission medium independently of the physical characteristics of the medium, but taking into account the topological aspects of the network, in order to enable the exchange of data between data station
MAN	Metropolitan area network (802.6 Standard)
MAP	Manufacturing automation protocol
MAU	Multistation access unit (Alternatively MAU or 8228)
MES	MAP end systems
MHS	Message handling system
MMFS	Manufacturing message format standard
MMS	Manufacturing messaging service
MODBUS	Modicon LAN product
MOTIS	Message oriented text interchange system (X.400 superset)
MRP	Materials requirements planning
MTA	Message transfer agents
MTS	Message transfer service
MUGCC	MAP/TOP User's Group Co-ordinating Committee
NBS	National Bureau of Standards
NC	Numerical control
NCC	National Computing Centre (UK)
NCC	Network control computer
NEMA	National Electric Manufacturers Association
NETBIOS	Network basic input/output system

Abbreviations

NETC	Network Evaluation and Test Center (ITI, Ann Arbor, Michigan)
NISL	Network interface sublayer (NISL provides a mapping function between the facilities of actual communications network (i.e. a LAN) and the services required of Layer 4).
NIU	Network interface unit
NMTBA	National Machine Tool Builders Association (USA)
NPAI	Network protocol address information
NRM	Normal response mode (HDLC Specific)
NSAP	Network service access point (The means of communication between the network service component and the end-to-end service component)
NSDU	Network service data unit
NT1	Network termination point number one
NTIS	National Technical Information Service
NTT	Nippon Telephone and Telegraph (Japan)
NTU	Network interface unit
OCTET	A collection of eight bits. Used instead of byte because it connotes eight bits
ODA	Office document architecture
OEM	Original equipment manufacturer
OSI	Open Systems Interconnection
OSIRM	Open systems interconnection reference model
PABX	Private automatic branch exchange
PACKETS	PDUs; Strings of bytes or octets represented in hexadecimal form
PAD	Packet assembler/disassembler (Provides packet switched network access for synchronous terminals)
PBX	Private branch exchanges. A manual or dial exchange connected to the public telephone network, operated by the user and located on his premises.
PCIF	Programmable controller interface
PCLNS	Protocol for connectionless-mode network service
PCM	Pulse code modulation
PCTE	Portable common tool environment
PD	Programmable device (In the MAP document this term is used in reference to plant floor devices, such as programmable logic controllers, robots, CNC Machines, etc)
PDN	Public data network
PDU	Protocol data units
PERA	Production Engineering Research Association
PGI	Parameter group identifier
PI	Parameter identifier
PLANET	Private local area network
PLC	Programmable controllers
PM	Phase modulation
PMX	Packet multiplexer
PROWAY	Process Data Highway
PSI	Programmable serial interface (HP term)
PSTN	Public switched telephone network
PTT	Postal, Telegraph and Telephone
RAS	Reliability, availability, serviceability
RCS	Reflex control system
RF	Radio frequency
RIA	Robotic Industries Association
RNR	Receiver not ready (HDLC Specific)

RS232C	A common interface standard that permits DTEs and DCEs to successfully interconnect.
RS499	An interface standard that enhances the capabilities defined by RS232C.
SA	Source address
SAP	Service access point
SASE	Specific application service element
SDLC	Synchronous data link communication (IBM bit serial link level protocol)
SDU	System data unit
SEAP	Service element access point
SLD	Subscriber-line datalink
SMF	Standard message format
SMUG	Swedish MAP Users Group
S/N	Signal-to-noise patio
SNA	Systems network architecture (IBM network protocol)
SNACF	Subnetwork access facility
SNDCF	Subnetwork Dependent Convergence Facility (intra-network), used to adjust the services provided by the network layer either upward or downward for interworking
SNDCP	Subnetwork dependent convergence protocol
SNICF	Subnetwork independent convergence facility (harmonizing)
SNPA	Subnetwork point of attachment
SONET	Synchronous optical network
SPA	Serial port application
SPAG	Standard Promotion and Applications Group (Brussels)
SPDU	Session protocol data unit
SSAP	Session service access point
STAR	Type of tree network in which there is exactly one intermediate node
STDM	Statistical Time-Division Multiplexer. An 'intelligent' time division multiplexer that uses microprocessors and memory to multiplex data of active channels only.
SVC	Switched virtual circuit
TAG	Technical Advisory Group
TBC	Token bus controller
TCM	Time-compression multiplexing
TCP/IP	Transmission control protocol/Internet protocol
TDM	Time division multiplexing
TIM	Token/net interface module (Concord Data Systems)
TNC	The Networking Centre (UK)
TOKEN	A symbol passed among data stations to indicate the station temporarily in control of the transmissions medium
TOP	Technical and Office Protocols. TOP, and the TOP logo are trademarks of the Boeing Company
TPDU	Transport protocol data unit
TRC	Technical Review Committee
TSAP	Transport service access point
TSDU	Transport service data unit
UA	User agent
UDLC	Universal datalink control. A Sperry Univac synchronous protocol
UI	Unnumbered information
ULI	Upper layer interface
VRC	Vertical redundancy check sum

Abbreviations

VT	Virtual terminal
VTP	Virtual terminal protocol
V.35	A CCITT interface recommendation that uses balanced circuit techniques for transmission of high data rates.
WAN	Wide area network
WD	Working draft, ISO standard status
WIP	Work-in-progress
XID	Exchange ID (IEEE 802)
XNS	Xerox Network Systems (protocols developed for the early Ethernet system produced by DEC, Intel and Xerox – still popular in the USA)
XPC	X.25 protocol controller
X.25	CCITT communications recommendation defining the connection of a computer ot a packet-switched network.
X.409	CCITT message-handling syntax
X3S3.3	ANSI committee working on network addressing
X3T9.5	ANSI committee working on the Fiber Distributed Data Interface (FDDI) standard.

SOURCES AND RECOMMENDATIONS FOR FURTHER READING

This book has drawn heavily on many documents. So many, if press releases and brochures of various kinds are included, that it is not possible to name them all. It goes without saying that General Motors and the US and European MAP and TOP User Groups have produced a wealth of information on the subject, including of course the MAP 2.2 and 3.0 and TOP 3.0 draft specifications. But some other documents have also proved particularly valuable. They include:

The 'Leeds report' by Carl Clarke and Ian Davidson on the setting up of a conformance test centre;
Various Motorola articles, including 'Carrier bands get nod for industrial networks' by Tom Ralph, 'Low cost networking for islands of automation' by Mary Gallagher, 'Systems integrators' MAP to promised LAN' by James Langdal, 'Token bus controller uses monolithic structure to chip away MAP costs' by Tom Balph, Rhonda Dirvin and Yehuda Shvager, 'The MAP program, an overview' by Tom Balph and Jim Vittera, plus any number of Motorola data sheets and manuals without which this book could not have been written;
The CNMA Implementation Guide and the documents relating to it, including a number of papers written at British Aerospace, were an invaluable fund of detailed information about MMS and many other mysteries.

Useful books included:

ICL's four-volume *The principles of Data Communications* edited by Geoffrey Carter, published by Heinemann Computers in Education, a key source of information about the seven layer model;
Computer Networks, Bacon, Stokes and Bacon, Chartwell-Bratt;

188 Sources and Recommendations for Further Reading

Fundamentals of Modern Digital Systems, Bannister and Whitehead, Macmillan;
CADCAM, Groover and Zimmers, Prentice-Hall;
Interactive Graphics in CAD, Gardan and Lucas, Kogan Page, 1984;
Intelligent Instrumentation, George Barney, Prentice-Hall;
Transducers for Microprocessor Systems, J.C. Cluley, Macmillan.

Other useful documents include:

The CIMAP Event Guide, published by Findlay Publications;
'Through MAP to CIM', by Eric Morgan, published by the DTI;
'Feeder Systems for the data highways of the future', paper by Harold Harrington of Sperry, now Unisys, UK;
'Local Area Networks,' paper by Cornell Drentea of Honeywell;
'Background on Industrial Automation', Intel;
'GM's Manufacturing Automation Protocol', Intel;
'An introduction to layered protocols', by Michael Witt of Micro Five Corporation, *Byte* magazine, September 1983;
'The CAP Guide to Open Systems Interconnection', a wallchart published by CAP Industry.

Bibliography

Anon
Computer Networks
Engineering, May 1986

Anon
Computers in Manufacturing Industry
Institution of Mechanical Engineers, London 1986

Anon
Electronic Data Systems (EDS)
Engineering and Manufacturing Group
Devonshire House, Mayfair Place, London W1X 5FH

Anon
Industrial Networking Incorporated (INI)
Belmont Road, Maidenhead, Berkshire SL6 6ND.

Anon
Manufacturing Systems Group,
Hewlett-Packard Ltd, King Street Lane,
Winnersh, Wokingham RG11 5AR

Anon
Motorola
Motorola MAP Design Seminar

Blanchar, D
Using the Manufacturing Automation Protocol (MAP) in the Factory
Hewlett-Packard Measurement Software Conference,
April 10 & 11 1986, CUPERTINO, CA

The Boeing Company
Technical and Office Protocols
Specification Version 3.0, Second Printing, November 1985, USA

Clarke, C.G. and Davidson, I
A Report to Define the Requirements and Assess the Benefits of a Comprehensive MAP Conformance Testing Service in the UK
Department of Trade and Industry

Deadman, R
Everything you always wanted to know about MAP
Industrial Computing, pp. 29-30, 1986

Douglas, P
The LAN Standard
Systems International, pp 59-60, Dec 1986

Durham, T
General Motors MAP Users Group
Society of Manufacturing Engineers
1 SME Drive, PO Box 930, Dearborn, Michigan 48121, USA

ESPRIT Project 955
Communications for CIM an Overview of the CNMA Project
Hanover Fair, Apr 1987

ESPRIT Project 95
Initial Conclusions of the CNMA Project
Hanover Fair, Apr 1987

Finnie, G
Meeting of Minds on Industrial LANS
Communications Systems Worldwide, pp 43-45, Dec/Jan 1987

General Motors
Manufacturing Automation Protocol – A Communications Network Protocol for Open Systems Interconnection
Versions 2.1.A & 2.2, 1 August 1986, USA

Goulding, M
Communications for CIM – A User View
Document D8, British Aerospace, England, 1986

190 Sources and Recommendations for Further Reading

Hansen P and Brenna, R
Distributed Intelligence in the Manufacturing Architecture
Hewlett-Packard Co, AMS, Presented at Automotive Automation Society Conference 1986

McCaig, M and Roden, R
A Matter of Protocol
Systems International, Dec 1986

Manchester, P
Lowering the Open Standard
Informatics, pp 12-14, Jan 1987

Merry, P and Heeler, A
CIM – the Communications Requirements
Logica (UK) Ltd, (Paper Ref. PM1AAG)

Moir, I
Managing Data Networks on an International Scale
Communications Systems Worldwide, pp 37-41, Dec/Jan 1987

Mottram, D.P.
Introduction to the manufacturing message service (MMS)
AMTEX 87 Conference, 8-11 September 1987, Telford, UK

Mulqueen, J.T.
LAN Industry Booms amid Fear of Coming Shakeout
Data Communications, pp 109-121, Mar 1987

One Woman's Vision of Open Systems Future
Communications Systems Worldwide, pp 35, Dec/Jan 1987

Pollard, T
Low Cost Links
Systems International, pp 65, 1986

Purdue, D.R.
Data Transmission Techniques for OSI/MAP/TOP Layer 1 – Physical Communications,
ComCentre, UK, Spring 1987

Purdue, D.R.
MAP 3.0 and TOP 3.0 – The Similarities and Differences
AMTEX 87 Conference, 8-11 September 1987, Telford, UK

Rose, C
BT lays its Plans for the Retail Section
Network, pp 23-26, Mar 1987

Scott Currie, W
The LAN Jungle Book – A Survey of Local Area Networks
Edinburgh Regional Computing Centre, 2nd Edition, Oct 1986

INDEX

aerospace industry, MAP communications in, 156-67
Allen-Bradley, 18, 101
American National Standards Institute (ANSI), 34, 98
amplitude modulation, 41, 44
application layer, 53
assembly information management system (AIMS), 133
  engine assembly case study, 133-9
  functions, 135-6
  solution to problems, 136-7
association control service element (ACSE), 54
automatic cooperation (Honeywell), 138
automatic guided vehicle (AGV), integration of, 18
automatic test equipment (ATE), 128

bandwidth, 32, 41
  Tandem demonstration, 118
binary oriented protocol, 71
Boeing, and TOP, 1, 3
  network architecture, 85
British Aerospace, 13, 38, 151-63
  FMS system, 153, 158, 160
    assembly procedure, 155-7
    communications in, 157
    justification for, 162
    tool preparation, 161
British Standards Institute (BSI), 97
broadband, 32, 41, 78
  and baseband, 41
  cable, 43
  coaxial system, 44
bus network, 32
bus and ring network topologies, 31

CAD system, 17
carrier band system, 73-6
carrier sense, multiple access, *see* CSMA/CD
cell control applications (Gould), 119 *et seq*
character oriented protocol, 71
client service agent (CSA), 60
CNMA project, 58, 104, 105, 151
coaxial cable transmission, 29
  transmission ratio, 32
  types in use, 33
collision
  detection, 36
  delays caused by, 37

common applications service element (CASE), 54, 147
communications, 23 *et seq*
  system, 20
companion standards, 107
Computer and Business Equipment Manufacturers Association (CBEMA), 98
Computer Communications Industry Association (CCIA), 98
computer integrated manufacture (CIM), 14
computer numerical control (CNC), 15
connectionless communication, 63, 64, 106
connectionless network protocol, 70, 106, 109
connectionless network service (CLNS), 69
connection-oriented communication, 63
  and multiplexing, 64
coordinate measuring machine (CMM), 17
Corporation for Open Systems (COS), 98, 112
CSMA/CD network, 36, 37, 38
Cummins AIMS system, 133
current loop, 80
  system, 29

dataflow and message traffic, 144
datalink layer, 70
  protocols, 71
design-assembly-inspection link, case study, 124
  and CAD/CAM, 124
  design, 124
  inspection, 126
  software, 126
design department, integration of, 19
  use of database, 20
Digital Equipment Corp (DEC), 3, 38
direct numerical control (DNC), 16
directory services, 59
distributed LANs, 30 *et seq*
dots, gold and green, 113
DuPont Corp, and MAP, 9

Eastman Kodak Corp, use of MAP, 9
economic batch quantity (EBQ), effect on, 19
Electronic Industries Association (EIA), 28, 102
electronic test and repair, and MAP, study, 127-32
  communication system, 132
enhanced performance architecture (EPA), 11, 76
Enterprise Network Event, 5, 113

error correction, 66
ESPRIT programme, 9, 151
Ethernet, 1, 35 et seq
　advantages of, 37
　contention network, 36
　CSMA network, 36, 37, 38
　traffic, 38
European MAP Users Group (EMUG), 9
European Standards Coordinating Committee, 98

factory communications system, 1
　Ethernet, 1
　RS232, 1
　TOP, 1
factory, use of MAP in, case study, 139 et seq
　CASE, 142
　dataflow and message traffic, 144
　Hewlett-Packard, AMSO, 139-48
　INI interface, 142
　MAP network access method, 142
　MMFS, 142
　PCIF application, 144
Ferranti (UK), 34, 79
fibre distributed data interface (FDDI), 34
fibre optics, 33 et seq
　network topologies, 35
field bus, 79
file transfer and access management (FTAM), 55 et seq, 101
flexibility, Tandem demonstration, 119
frame length, 72
frequency division multiplexing (FDM), 42
frequency shift key (FSK), 41, 74

gateway, 81 et seq
General Motors, MAP, 2, 163
　changeover time, 160
　communication system, options, 3, 25
　'factory of the future', Saginaw, 5, 10, 163
　GM10, 9
　GMT400 truck and bus project, 9, 167
　unmanned plant, 165
　Vanguard project, 163-7
Gould cell controller (GCC), 119
　applications, 121, 122
　benefits, 121, 123
　configuration, 121, 123
　problems, 121, 123

head ends, 44 et seq
high level datalink control (HDLC), 73

IBM, 19, 139
ICL (UK), 9, 78, 124-7
　case study, 148-51
incompatibility problem, 11
industrial automation market, size of, 13

industrial networking, 146
information flow dynamics (Honeywell-Bull), 137
INI interface, 142
initial graphical exchange specification (IGES), limitations of, 25
injection laser diode (ILD), 33
Instrument Society of America (ISA), 73, 80, 99, 107
Institute of Electrical and Electronic Engineers (USA), 35-6, 39, 73-6, 96-7, 99
integrated service digital network (ISDN), 112
intermediate systems, 83, 84
International Standards Organisation (ISO), 28, 61, 91, 92
　secretariat, 93
interoperability of systems, 109
interpersonal messaging (IPM), 60

Jaguar Cars, 9, 17
job transfer and manipulation (JTM), 59

Kaiser Aluminium & Chemical Corp, 9

layers,
　communication between, 52
　control information in, 52
　data, passing of, 52
　datalink, 71
　grouping of, 48
　modes, 47 et seq
　network, 69
　physical, 71
　presentation, 62
　protocol, 62
　session, 62
　transport, 63, 65
light emitting diode (LED), 34
local area network (LAN), 13
logical link control, 71

manufacturing message format standard (MMFS, Memphis), 10, 56, 101, 142
manufacturing message service (MMS), 56-7, 101-9, 113, 153
　ambiguities, 107
　coding, 58
　file transfer mechanism, 104
　and MAP specification, 103
manufacturing resource planning (MRP1, MRP2), 14
medium access control, 71
Memphis, 101
message handling system (MHS), 60
message transfer system (MTS), 60
migration, 106
miniMAP, 76

Index

modem, 27
multiplexing, 64
  transport layer, 65
network layer (layer 3), 69
  structure, 69-70
network management, 61-2

office documentation architecture (OPA), 89
off-line programming systems, 15
open systems interconnection (OSI), 4, 47
  origins, 91 *et seq*

PCIF applications, 144
phase shift keying CAM/PSK, 44
physical layer (layer 1), 71
presentation layer (layer 6), 62
printed circuit board assembly, use of MAP in, 148-51
  broadband coaxial cable, use of, 150
  development, 148
profibus (process field bus), 81
programmable controller (PLC), 17
project 955, ESPRIT, 152
proprietary mail system (PRMD), 61
protocol data unit (PDU), 51
Proway, 9, 73

quality control data, 17

real time response (Honeywell Bull), 138
reliability, Tandem demonstration, 118
repeater, 9, 84
ring network, 32
robot program, 18
Rover Group, 9
RS232 communications standard, 27-30, 42
  rate of operation, 28
RS422, 423, 449, 488, 27-30
RS511, 102-4
  see also MMS

SASE, 54, 59
session layer (layer 5), 62
seven layer model, 47 *et seq*,
  and addressing, 48
  open system interconnect model (OSI), 47

standard message format, 101
standards, development of, 95
  draft international standard, 96
  draft proposal, 96
  international standard, 96
  new work proposed, 95
  working draft, 96
star network, 30
statistical process control, 17, 130
synchronous optimal network (SONET), 34

Tandem, MAP demonstration, 116
tape reader system, 15
technical and office protocols (TOP), 1, 55, 59-62, 84 *et seq*
  3.0 specification, 88
testing of MAP, 109
  conformance, 111
  error recovery in, 111
  interoperability, 111
  test tools, 110
Texas Instruments, 18, 101
time division multiplexing (TDM), 43
token passing, 38 *et seq*
  access time, 39
  bus or ring, 39
  Cambridge ring, 41
  error recovery, 40
  slotted ring, 41
'Towers of Hanoi' puzzle, MAP, 115
transport layer (layer 4), 63 *et seq*
transport service access point (TSAP), 65
twisted pair wiring, 26
  computer data transmission, 26
  connection method, 27
  limitations, 27
  other systems, 29 *et seq*
  rate of operation, 27

UK, MAP activities in, 97
user agent (UA), 60
US MAP/TOP Users Group, 5

virtual terminal (VT), 59
vision system, and MAP, 116

wide area network (WAN), 66-8